Simple and Fun™ SCIENCE SIMPLIFIED

Learning by Doing

National Standards Included

Authors
Dennis McKee
Lynn Wicker

Illustrations
Dennis McKee

Design
Jeff George

Senior Consultant
Mark Gilbert Twiest, Ph.D.
Associate Professor, Science Education
Indiana University of Pennsylvania

Consultants
Michael B. Leyden, Ed.D.
Todd Fennimore, Ph.D.

P.O. Box 2590 • Columbus, OH 43216-2590

Many of the activities contained in this book first appeared in *Teaching PreK–8* magazine.

Printed in the United States of America
10 9 8 7 6 5 4 3 2 1

Introduction to Simple and Fun Science

Science is a part of everything we do and everything we know. Young children have a natural curiosity about their world and how it works. This is science! As children grow older and their knowledge base expands, so too does their curiosity—but only if this interest is fostered and encouraged.

Educational researchers now know the critical role that attitudes play in learning all subject areas. Students who enjoy an activity or topic are much more likely to learn more about that subject. So how do you foster this curiosity? One of the most important things you can do is to display a positive attitude yourself. Exploring your world stops for most busy adults, but if you can rekindle this curiosity yourself and model positive attitudes toward science, your child is much more likely to remain interested as well. Getting involved in the science activities in this book will help your family maintain and grow the positive attitudes that will lead to a greater understanding of the science that is all around us.

Too many times we think of science as a subject that we study only in school and one that doesn't really affect us. However, the fact is that all of us use the same kinds of skills that scientists use every day. These skills, called process skills by science educators, include things like

- observation
- communication
- prediction
- measurement
- inference
- hypothesizing
- identifying and controlling variables
- experimenting
- drawing conclusions

As we engage in these processes, the only difference between a scientist and the rest of us is the level at which we use these skills. For example, when a scientist observes something, he or she is likely to use very specific language to describe the object and describe it with both words (a qualitative description) and numbers (a quantitative description). In addition, scientists will use all of their senses in describing this object. They may make repeated observations and then compare their observations with those made by other individuals.

It is this level of accuracy in description that makes the observations of a scientist very different from those of a layperson. The same is true with the other process skills as well. This book will help you learn to use the process skills in a way that a scientist would and have fun learning important science facts along the way.

Science educators acknowledge the importance of each of these three areas in science classrooms for students of all ages. Positive attitudes, a working knowledge of the process skills, and science content together give our students the opportunity to excel in science and become the lifelong learners that every parent and every educator hopes they will be.

This book is intended for young children seven to nine years of age. It is important that you help your child by guiding him or her through the process and not simply doing the investigation yourself. Encouragement and positive reinforcement are the key in this process. You may in some cases wish to participate by doing a similar activity and then comparing results.

Don't let your child become discouraged if the results are not as anticipated or predicted. Many times in science the most interesting things happen and the greatest learning takes place when the unexpected happens! Encourage questions and do not be afraid to say "I do not know" and instead reply with "I wonder how we could find out."

In this information age, the ability to supplement the activities in this book is as close as the library, one of the many science-related channels on the television, or the Internet. Local schools, universities, museums, planetariums, aquariums, and nature centers also have abundant resources and can provide answers to questions that arise from the activities your child will try. Remember, the most important thing you can do to help your child succeed is to foster positive attitudes toward science followed by learning the process skills. The science content is sure to follow.

Mark Gilbert Twiest, Ph.D.
Associate Professor, Science Education
Indiana University of Pennsylvania

National Science Standards

1. Unifying concepts and processes in science
2. Science is inquiry
3. Physical science
4. Life sciences
5. Earth and space science
6. Science and technology
7. Science is personal and social perspective
8. History and nature of science

Simple and Fun Science
Book D • Ages 7–9

Page	Lesson Title	National Science Standard
10	Describe Living and Non-Living Things	1, 2, 3, 4, 5, 7
10	Plants Are Living Things	1, 2, 4
11	Tracking Wild Plants	1, 2, 4
12	Plants Are Our Foundation/Testing a Hypothesis: All Plants Need water. All Plants Need Light	1, 2, 4, 7
13	Two Types of Plants	1, 2, 4, 7
13	Helpful Plants	1, 2, 4, 7
14	Male and Female Parts/Look Closer at the Plant and Its Flower	1, 2, 4
15	Plant Needs	1, 2, 4
16	Two Types of Trees	1, 2, 4, 7
16	Tree Spider Map	1, 2, 4, 7
17	Fruits	1, 2, 4
17	Seeds	1, 2, 4, 7
18	Identifying Foods from Plants	1, 2, 4, 7
19	Can Seeds Move?/Simulate a Maple Tree Seed	1, 2, 4
20	Seeds Are Like Baby Plants/Inside a Seed/ Transplanting a Seedling	1, 2, 4
21	Animals Have Needs	1, 2, 4
22	Animals Come in All Shapes and Sizes	1,2,4
22–23	Bats/See Like a Bat/Sound Travels Through Vibrations	1, 2, 3, 4, 6, 7
24	Hunting for Food/Simulate an Animal That Uses Its Tongue to Get Food	1, 2 ,4
24–25	Birds' Beaks/Like a Hummingbird Beak Chopsticks Are Like Birds' Beaks	1, 2, 4, 7
25	A Jungle Jingle	1, 2, 4, 7
26	Animal Movement	1, 2, 4, 7
27	Skeletal and Muscular Systems	2, 4
28	How Many Bones and Muscles Do You Have?/Muscles Relax, Too	1, 2, 4
28	Muscles and Bones Are Spectacular Machines	1, 2, 3, 4, 7
29	Your Head and Neck at Work/ You Can Make Your Muscles Work	1, 2, 3, 4, 7
30–31	Your Elbow, the Fulcrum/Make a Lever Like Your Arm/ Thigh Bone Connected to the Leg Bone	1, 2, 3, 4, 7
31	Cartilage Reduces Friction	1, 2, 3 ,4

National Science Standards lesson assessment by Cynthia S. Snyder, 5th-Grade Teacher, Lake Elementary School, Western Wayne School District, Lake Ariel, Pennsylvania

What to Do as You Work Through This Book

Make sure children:

1. Pay attention to the work from start to finish.
2. Clean equipment before and after an activity.
   Keep things clean and put things away when finished.
3. Label containers in experiments.
4. Work with a partner or an adult helper. Scientists usually work in teams.
5. Think about safety first. Here are some tips:
   Never look directly at the sun.
   Protect your eyes from anything that can splatter or shatter.
   Never get close to animals in the wild or even strange pets.
   Never taste anything unless you know exactly what it is and that it is safe.
   Stay away from fire, burners, very hot water, or anything that can burn.
   Be very careful with scissors and ask an adult to help when using any cutting tool.
6. Keep a science response journal. Write down questions when they arise. Write down observations. Interpret and represent explanations of concepts and new information in their own words. Draw pictures of observations. Design experiments and new investigations.
7. Take time. Try not to hurry. Try not to get frustrated. If something doesn't work the first time, try again. Everyone makes mistakes. Sometimes failures in science raise new questions and uncover new solutions. And sometimes solutions to problems cause new problems requiring new solutions.
8. Stay organized.

Before you and/or children start any activity or experiment:
Read through the activity first.
Gather all the materials that will be needed for any activity.
Follow each step. Don't skip steps.

As children do an activity or experiment guide them to:
Be very accurate.
Be very careful and work cautiously.
Be very neat so the work and the work area are neat and clean.
Be creative.

Essential Learning Products *Simple and Fun™ Science Simplified* series encourages children to work at their own pace. Book D is intended for children from ages seven through nine. Use it as a supplement to existing curricula in the classroom or independently at home. Try not to impose your own conclusions on the children. Stand back and allow the child to find out. However, children do need your guidance.

Table Of Contents

Asking Questions, Finding Answers, and Solving Problems

Why? How many? When? Are you sure? Keep asking questions. That's what scientists have done since the beginning of time. How can we keep warm? Why does the sun seem to disappear from the sky at the end of the day? What keeps my heart beating? How deep can we explore in the oceans? How far into space will we travel? Can we build a space colony to support future generations?

Scientists work to answer questions and to solve problems. Scientists cannot answer every question. And scientists often disagree. But scientists do try to explain, predict, make sense of the world around us, and solve problems.

Most Scientists Work Through a Problem-Solving Process

They don't always follow these steps exactly the same way every time they look for answers and solutions to problems, but they usually do each step.

1. Define the problem: What am I trying to find out?
2. Gather evidence: What do I already know? What do I need to find out?
3. Make a prediction: What do I think will happen?
4. Experiment: How can I test my prediction?
5. Gather results: What did my experiment show?
6. Draw a conclusion: What did I learn about the problem? Did I find possible ways to solve the problem?
7. Raise new questions: What do I still not understand or what new problem does my conclusion cause?

Remember these steps as you work through this book. Start a science journal. Remember to write down questions and predictions. Write down the information and evidence you gather. Make sure you keep good records of what you do and what you observe.

Describing Living and Non-Living Things

What are some characteristics of living things that you already know?

____

List some living things you know are living and how you know they are living.

____

List some non-living things around you. How do you know they are non-living?

____

All sorts of living things are all around you. Look outside. Do you see animals? Do you have pets? A dog maybe? A cat or a goldfish? Maybe a bird? Do you see a squirrel or a bird outside your window? Your dog or cat, the squirrel, and any bird you see are animals. What about a spider or an ant? Do you see any spiders or ants around? What about a fly? They're all animals, too.

Plants Are Living Things

What about plants? Of course, they're living things, too. Do you see any trees around you? What about flowers? Do you have green plants in your house? Plants and animals are living things. First, let's learn a little about plants. Then we will start learning about animals.

Do you have any questions about plants? What are they?
Write them here or in your science journal.

____

As you learn about plants, try to answer some of these questions, too.

What are some different kinds of plants? ______________________________

____

What parts do plants have? ______________________________

What do plants need to live? ______________________________

How do plants change during different seasons? ______________________________

____

What are seeds and what are they for? ______________________________

____

Tracking Wild Plants

Plants are all around you. You only need to look out your window to see all kinds of plants that live in your area. Plants are extremely diverse. What do you think *diverse* means? Try to find out. Here's a suggestion: Get a good dictionary and keep it with you as you work through this book. Sometimes new words might slip in without a definition. If they do, find out what the word means, then read the sentence again.

Plants are so diverse because of the numbers of shapes, sizes, and colors. In the fall, the colors in some areas are brilliant with the reds, yellows, and oranges of the leaves. In the spring, summer, and fall, many areas are also vibrant with the pinks, purples, yellows, reds, and even blues of different flowers. Write down some of the plants you can see from your window.

With an adult helper, take a walk around your neighborhood. If you or your adult helper have a camera, that is a wonderful way to record what you see. But it doesn't matter if you don't have a camera. Your eyes, a science journal, and a pencil can be your camera.

What you need:

a pencil
a nice day outside
a science journal
an adult helper

What to do:

1. Take a walk in your neighborhood or to a local park. Record the kinds and number of trees, flowers, shrubs, and weeds you see on your walk.
2. Record signs of plants, too. What are some signs that plants live in the area? Do you see any fruit? Do you see any twigs? What other signs might show plants are in your neighborhood or park?
3. Decide which plants you think are growing wild in your neighborhood or park. List them. Describe what makes you think they are growing wild.
4. Decide which plants you think are domestic plants or plants that were planted there on purpose. List them. Write what makes you think they were planted on purpose.

These are drawings of three poisonous plants to avoid. Watch out for them. Don't touch them.

Poison Sumac

Poison Ivy

Poison Oak

Plants Are Our Foundation

Plants are the living foundation of our planet. They absorb sunlight, and with carbon, they make food for themselves and for other living things in their environment. Without plants, many other living things, including you and me, could not survive. But plants have needs, too.

All plants need water. All plants need light. Let's start by experimenting to see if these two statements are true.

You're going to read a new word in the heading below. Do you have your dictionary? Before you look up the new word—*hypothesis*—write what you think it means. Try to figure out its meaning by how it is used here.

Hypothesis means ____

____

My Own Hypothesis About Plants:

____

____

Testing a Hypothesis: All Plants Need Water. All Plants Need Light.

What you need:

four plants that are alike
a cardboard box
paper
water
sunlight
pencil

What to do:

1. A label is a sign with words on it. Use the paper to make four labels: (1) WATER AND LIGHT, (2) WATER AND NO LIGHT, (3) NO WATER AND LIGHT, and (4) NO WATER AND NO LIGHT.

2. Place one label next to each plant. Do as the labels say for one week. Put a box over the plants that say NO LIGHT. But once a day lift the box. Give a little water to the plant that says WATER AND NO LIGHT.

3. After one week, look at all four plants.

4. Write in your science journal what each plant looks like.

Try to keep all four plants healthy. What do you think you can do? What are some other needs plants have besides water and light? Here's a hint: How long do you think your plants would survive without soil? What ideas did this experiment gave you about caring for plants?

Two Types of Plants

There are two main types of plants: flowering and non-flowering. Many of the non-flowering plants are the simplest kinds of plants, such as some kinds of algae. Some biologists call all algae protists, not even simple plants. The other simple non-flowering plants include conifers, mosses, and horsetails.

Most of the plants on the earth are flowering plants. There are 18,000 different kinds of orchids alone. It's hard to believe there are so many flowering plants because many, such as grasses and some trees, have flowers that are not easy to see. Duckweed floats on ponds and has a flower only $\frac{1}{100}$ of an inch big. Try to find something that measures only $\frac{1}{100}$ of an inch. Some trees, such as conifers, are non-flowering plants. Other trees, such as oaks and elms, are flowering.

Helpful Plants

A lot of plants are very helpful to us. Flowers are pretty and colorful for a very good reason. Their color attracts different insects and some birds that dine on the sweet nectar in the flower. Some flower nectar is used by the honeybee to make honey. The insect gets food from the flower, but the plant gets something out of the deal, too. Insects help the plants produce new plants.

Here's a "blooming" good idea. Make some fun cards and notepaper as presents. Then you can use your pressed flowers to study their parts.

What you need:

some fresh flowers
paper towel
notepaper
two heavy books
note cards
glue

What to do:

1. First, don't pick flowers without permission.
2. Lay the flowers on a piece of paper towel.
3. Lay a second piece of paper towel on top of the flowers.
4. Place two heavy books on top of the towels and flowers. Leave them for a week to press the flowers. When they are pressed, glue the flowers on note cards and paper as presents or to write letters to friends and family.

Male and Female Parts

Most flowers contain male and female parts. The anthers make something called pollen. The ovary contains the egg, sometimes called an ovule. For a new plant to be produced, the pollen and the ovule have to get together. It's hard for the plant to get them together at the right time by itself. That's where the insect helpers come in. Insects and sometimes other animals carry the pollen from plant to plant. This helps get the pollen and the ovule together so new plants will grow. Look at your pressed plants and see if you can find the different parts.

Look Closer at the Plant and Its Flower

What you need:

flowers from your craft activity
an adult helper
a piece of light-colored construction paper
a green plant with roots
a pencil
glue

What to do:

1. Use any flowers you used in the last activity. Ask an adult helper to help you cut the flower in half and mount it on light-colored construction paper.
2. Look at the flower in the picture. Compare the picture with your flower. Try to find the petal in your mounted flower. Try to find the anthers. Try to find the pollen on the anthers. Try to find the ovary. When you find them, label them on your mounted flower.

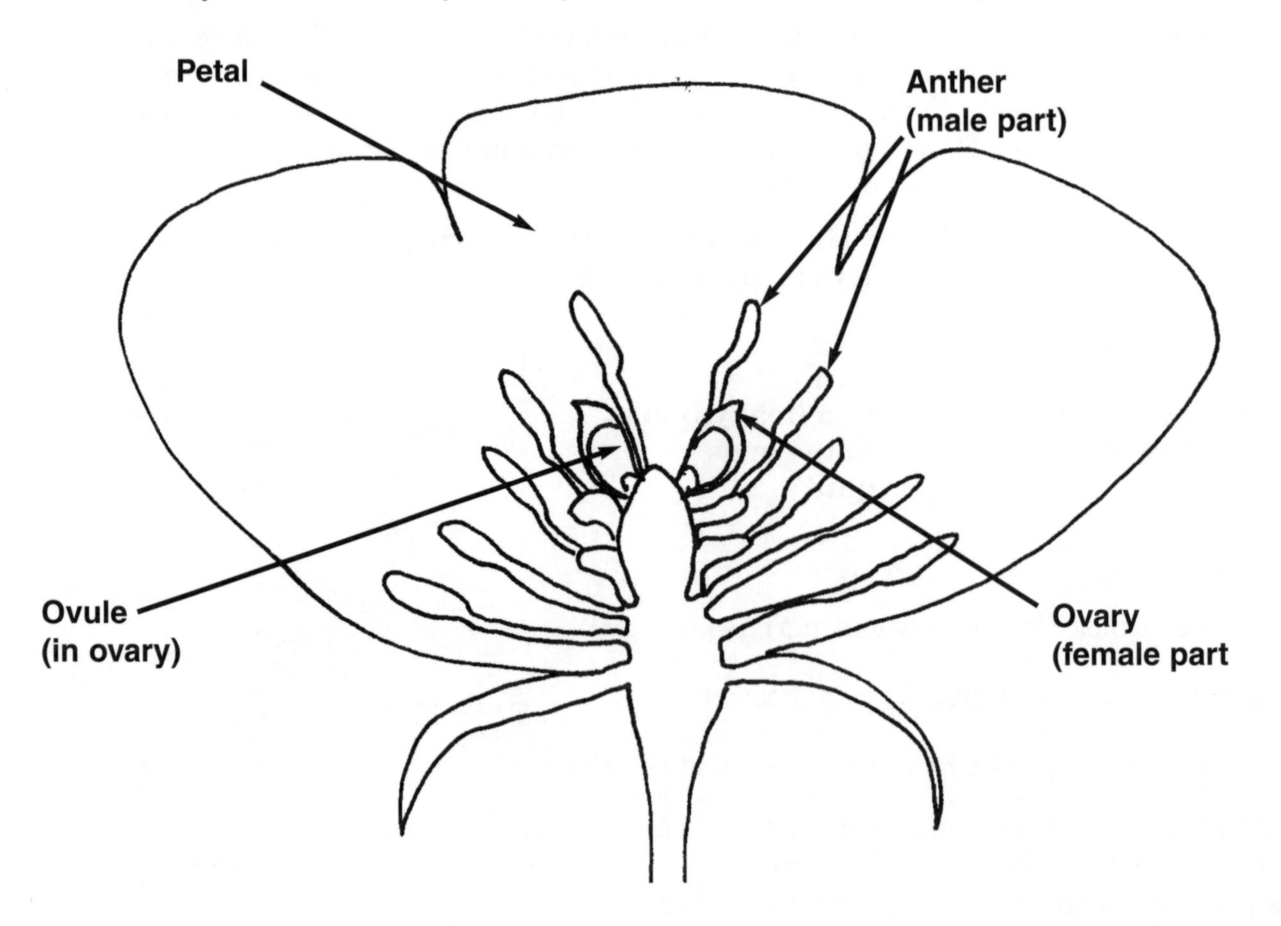

Plant Needs

Look at the plant in this picture. What helps it grow? What are some things you just learned plants need to grow? Plants need light. And like all living things, plants need food. Plants make their own food. You'll learn how they do that in another book. Let's look at a plant now and see what its other parts are.

What you need:

a plant with roots
a pencil
construction paper
glue

What to do:

1. If your flower has leaves, a stem, and roots, glue the entire plant on the construction paper. If not, ask an adult helper to help dig up a weed. If you can find a weed that is a flowering weed, that will work the best.
2. Glue your weed on a piece of construction paper.
3. Label the parts of your plant. Compare your flower and your plant with the plant in the picture.

How are the plants alike? ____________________

How are the plants different? ____________________

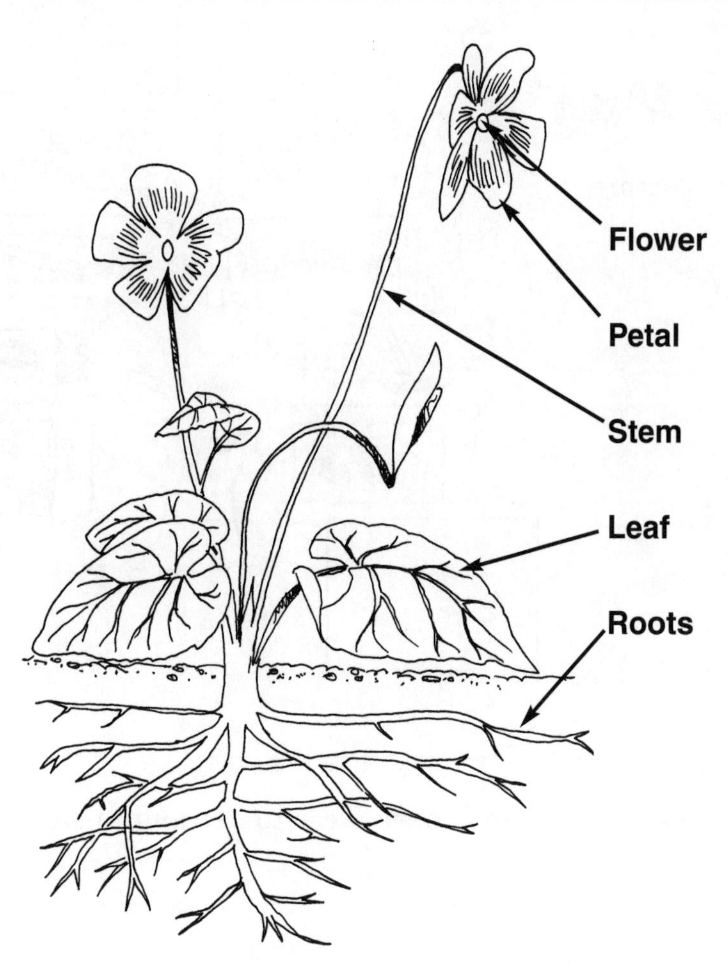

Two Types of Trees

There are two main types of trees on the earth: conifers such as pines and firs, and deciduous (broadleaf) trees such as oaks, maples, eucalyptus, and palm trees. Ginkgoes and cyads are trees, too—very rare trees. Actually, a tree is a tall land plant that has developed a very strong stem so it can grow tall to reach above other plants for sunlight. Redwood trees in California can grow to be well over 300 feet tall.

Trees are helpful to us, so we really need to be careful about conserving them. Now there's a big word, *conserving*. Try to find out what that means because there are a lot of animals, plants, and resources we need to conserve.

Tree Spider Map

Look at the spider map picture.

Tell how you think some trees are helpful.

Try to think of other ways trees can help us. Draw your own picture of the ways trees help.

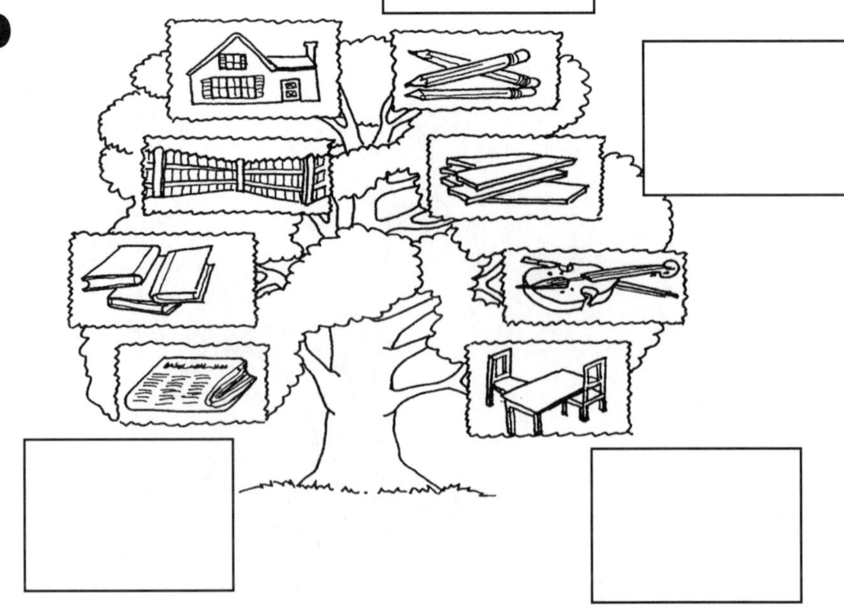

What do we use trees for?

____

How can we try to make sure we always have trees to use and to enjoy?

____

Fruits

Do you know what a fruit is? You probably think immediately of something like an apple, orange, cherry, or banana. Well, you're right. But scientists called botanists use the word *fruit* a little differently than you and I do. Botanists are scientists who study plants. To a botanist, a peanut is a fruit, too. So are tomatoes, green beans, and avocados. Once an ovary and pollen get together, the ovary in the flower begins to change into a fruit, a botanist's kind of fruit.

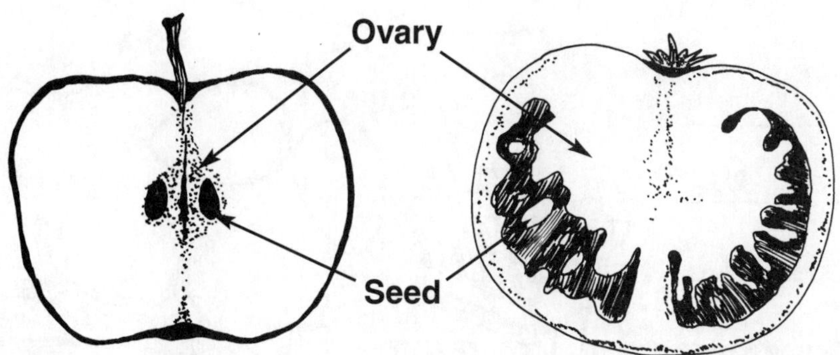

Find some other things a botanist would call a fruit. Save them for the next experiment.

Are the seeds always on the inside of the fruit? ______________________________

Where else might they be? ______________________________________

____

Seeds

Many seeds are found in the foods we eat. Ask an adult for help looking in the refrigerator for some foods that might have seeds in them. Together, you can look at different seeds and compare how they grow, how many you can find, and how different they are. You might want to do this around lunchtime so you can use the food you study and not waste it.

What you need:

some foods: apple, orange,
pickle, cucumber,
tomato, seeded grapes,
or strawberries
an adult helper

What to do:

1. Look at the food you find. Are seeds apparent without cutting into the food? Which food?
2. Ask your helper to cut a section of the food and look for seeds inside. Which foods have seeds inside? Draw what you see and describe each food and its seed or seeds in writing.
3. Cut up the foods and serve them to your helper as a snack.

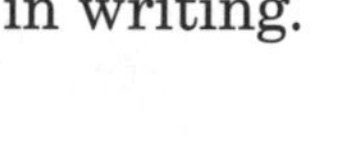

Identifying Foods from Plants

Look at the pictures of food.

Circle the foods you think come from plants.

Where do you think the other foods come from?

____

____

____

Most of these foods grow as they are on plants. Which of these foods are people-made? What are the ingredients? Where do you think the ingredients came from?

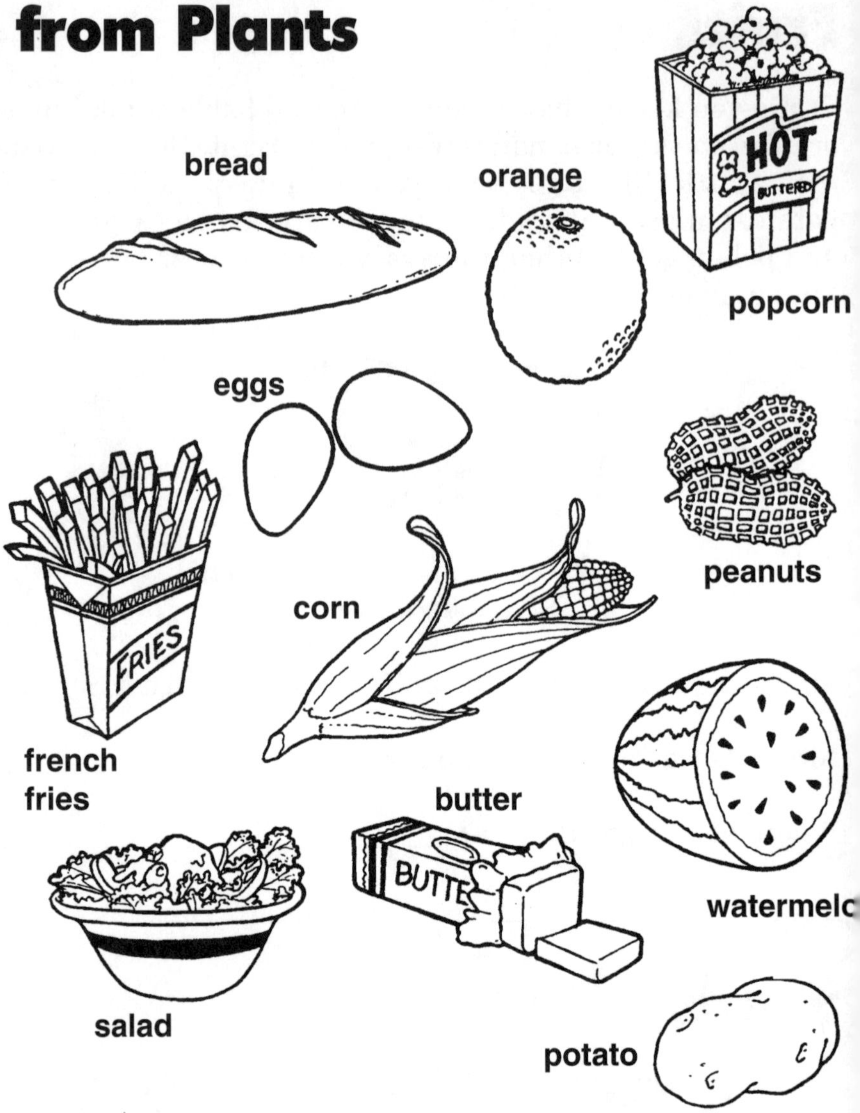

____

____

____

bread
orange
popcorn
eggs
peanuts
corn
french fries
butter
watermelo
salad
potato

List and draw some other foods you like that come from plants.

Can Seeds Move?

Many plants grow from seeds. Seeds, like plants, can be very different. Below is a picture of an acorn and how it grows into a giant oak tree after many, many years. Seeds can be blown by the wind, carried on animals, float down streams, and move in many other ways to new places to grow. Some seeds such as maple seeds are shaped like propeller blades so they can float through the air to new places to grow.

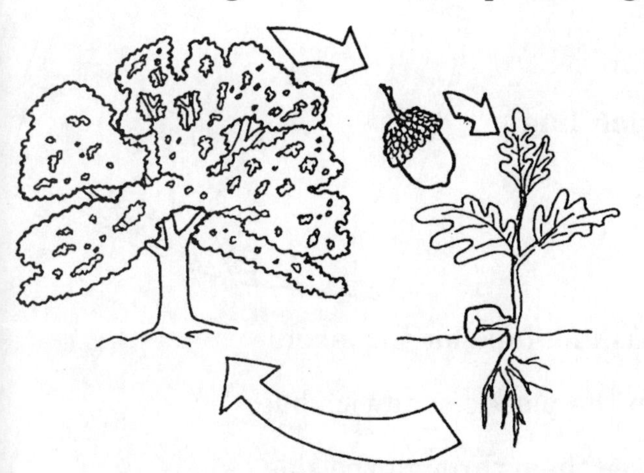

This is a pea plant, a pea pod, and a sprouting pea plant.

Simulate a Maple Tree Seed

You can make a model that floats a lot like a maple tree's seed floats.

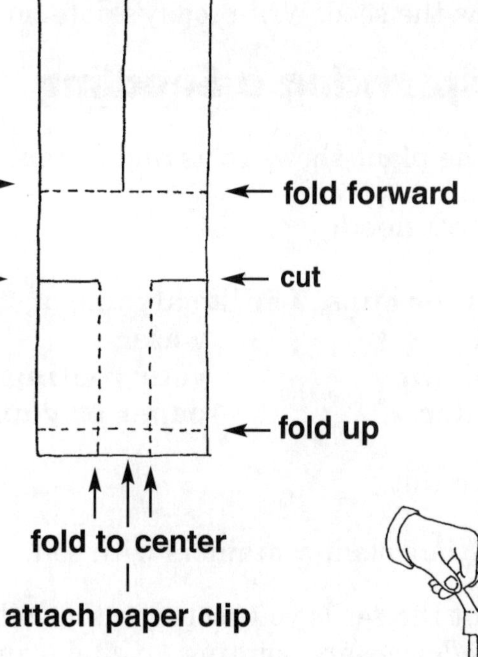

What you need:

construction paper
scissors
a pencil
a small paper clip

What to do:

1. Cut a piece of construction paper into long strips like the one shown here. You can use this as your model.
2. Cut one end down to the center and bend the two ears in opposite directions.
3. About one third up from the bottom, cut about one third in from the outside.
4. Fold the lower third into the center from each side and fold the bottom up about ¼ inch. Attach the paper clip to the bottom of your simulated seed.
5. Stand up holding the paper seed high in the air. Drop it.

Write about how it drops to the floor. Maple seeds are sometimes called whirlybirds.

Seeds Are Like Baby Plants

Some seeds use the food surrounding them as they grow roots and stems. You can watch a seed begin to grow. Then you can plant the seed you started and take care of it so it can grow into a strong, healthy plant.

Inside a Seed

What you need:

a bean or pea seed
cotton balls
masking tape
a plastic sandwich bag
water
a pen or marker

What to do:

1. Label your plastic bag with your name. Use masking tape for the label.
2. Dip the cotton balls in water and place them in the plastic sandwich bag.
3. Place the seeds on the cotton balls so you can see them through the side.
4. If you have a bulletin board, you can tack the sandwich bag to the board. Now watch the seed begin to grow a root and a stem. Record what you see. Draw the seed. Write today's date on the drawing.

Transplanting a Seedling

When the plant shows roots and a stem, you can transplant it into rich soil.

What you need:

2 paper cups, 2 milk cartons, or 2 flowerpots
soil
sunlight
a ruler
water
your rooting seeds
paper or your science journal

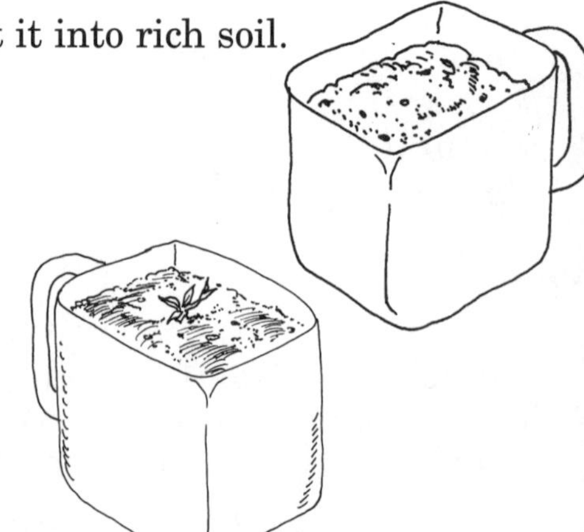

What to do:

1. Fill your plant containers with soil.
2. Plant the seeds you rooted in the soil so the root is pointing down and the stem/leaves are pointing up. Make sure your seed is just covered by dirt and the stem/leaves are sticking out of the soil. Water your new plant.
3. In a journal, record the date you started your seeds rooting and the day you planted them. Draw a picture of your little garden and how you planted the seeds.
4. Decide where to place your plants. Remember, what do they need to grow? Keep a record of their growth with drawings. Record information such as how often you water and the measurements of the plants' growth. Take good care of your new plants. You may want to study them later in your own experiment.

Animals Have Needs

Like plants, animals live in places where their needs are best satisfied. What do you think are some needs animals have?

___

What do you think the word *satisfy* means? How can you find out?

___

If animal needs are met, then they are suited to their special place or their environment. Animals have special structures or body parts that help them in their environment. Some animals can fly. Name some flying animals.

___

Some animals can only swim. Name some animals that can only swim.

___

___

Some animals can only walk, crawl, or run. Name some.

___

Name at least one animal that can crawl, walk, climb, and swim.

___

Name some animals that use their hands or paws as tools to find food.

___

Name some animals that use beaks as tools to find food.

___

Some animals have very good eyesight. Try to name some animals you think have very good eyesight for finding food.

___

Can you think of any animals that might not have good eyesight for finding food? Try to name one.

___

Animals Come in All Shapes and Sizes

What makes animals different and special also helps them find food, helps them eat and drink what they need, helps them protect themselves, helps them have baby animals, helps them move, and helps them in many other ways in their environment.

Bats

Some animals can't see with eyes, but they can still move safely and find food. Most bats send out sounds that bounce off objects and return to the bat's ears as echoes A bat can decide where objects are, how big objects are, even the shape of objects so they know what is food and what they might run into.

Sound travels through air, through water, and through solids. When you speak, sound vibrations travel from your mouth through the air. But sound vibrations also travel through the bones and body fluids through your neck and head. Let's try using vibrations the way a bat does for "seeing." Let's try to "see" what a bat "sees."

See Like a Bat

What you need:

a helper **blindfold or cover**

What to do:

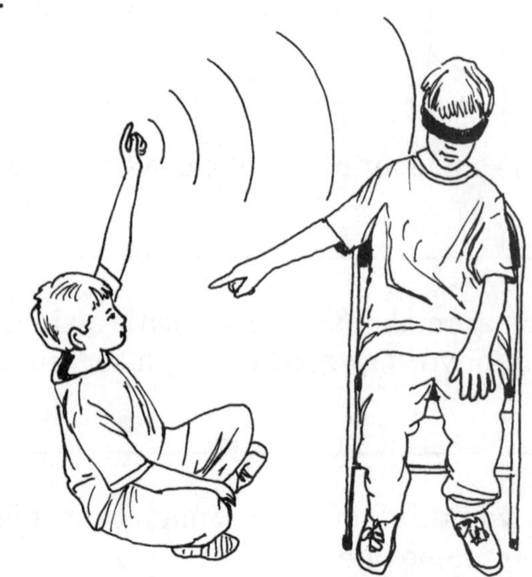

1. Close your eyes or put a cover over your eyes.
2. Ask your helper to click his/her fingers or slap his/her hands above you, behind you, to your right, to your left, below you, and directly in front of you.
3. Guess where the sound is coming from each time. Which direction was the hardest to figure out? Why do you think so? How do you think determining directions with your eyes closed simulates how a bat finds food?

Sound Travels Through Vibrations

Bats can tell what their prey is like by listening for changes in vibrations. Sound vibrations traveled through the air when your helper clicked fingers or clapped hands. But sound vibrations travel through liquids and solids, too.

What you need:

a string
one large metal spoon
two small metal spoons
a helper

What to do:

1. Tie the string to one of the small spoon handles so the spoon hangs in the middle of the string.
2. Wind one end of the string around one index finger. Wind the other end of the string around the other index finger.
3. Put your left finger lightly in the entrance of your left ear. Put your right finger lightly in the entrance of your right ear.
4. Lean forward so the spoon hangs loosely. Ask your helper to tap (lightly) with the other small spoon against the hanging spoon. Describe what you hear.
5. Tie the string in the same exact way to the large spoon and repeat step #4.
6. Now, do steps #4 and #5 again. But this time keep your eyes closed. Describe how you can tell the difference in sounds between the small spoon and the large spoon.

Describe how this experiment might show how bats and some other animals might find their prey.

____

Name some animals that you think use sound vibrations to find food and to move safely through their environment.

____

How can you find out more about these kinds of animals? Find out more and write what you find out in your science journal. How do scientists use sound to find things?

____

Here's a hint: What kind of scientist might work on a boat? How might they use sound vibrations to find things they can't see?

Suppose you had to find a stud in a wall to hang a picture. How do you think you could find one?

____

Hunting for Food

Animals have different tools for hunting food, too. Some use a long tongue to catch their food. What animal (or animals) can you think of that hunts with its tongue?

Simulate an Animal That Uses Its Tongue to Get Food

What you need:

a party whistle
masking tape
pieces of confetti
(paper-punch holes work well)

What to do:

1. Hold the party whistle in your mouth and get ready to blow into it. Throw some confetti up in front of you.
2. Blow into the party whistle and try to catch some confetti on the paper part of the whistle. What could you do to catch more? Remember, you have some masking tape.
3. Try modifying your whistle to see if you can catch more confetti. Describe what you did and how it worked. Draw a picture of your food-getting tool.

Birds' Beaks

Some birds use their beaks to get their food. Have you ever seen a hummingbird? Hummingbirds use their beaks in a special way to get their food from flowers. You use a tool sometimes in the same way a hummingbird uses its beak to suck out nectar from plants. What tool do you think that could be? Let's see.

Like a Hummingbird Beak

What you need:

a glass of milk **a straw**

What to do:

1. You predict what you should do to simulate a hummingbird getting nectar from a flower.
2. Describe what you did. What is your straw like in this picture? What is your glass like? What is the milk like?

Some birds use their beaks well for getting food for themselves and their babies and for other kinds of work, such as building nests. Have you ever used chopsticks? It takes practice to use chopsticks well.

Chopsticks Are Like Birds' Beaks

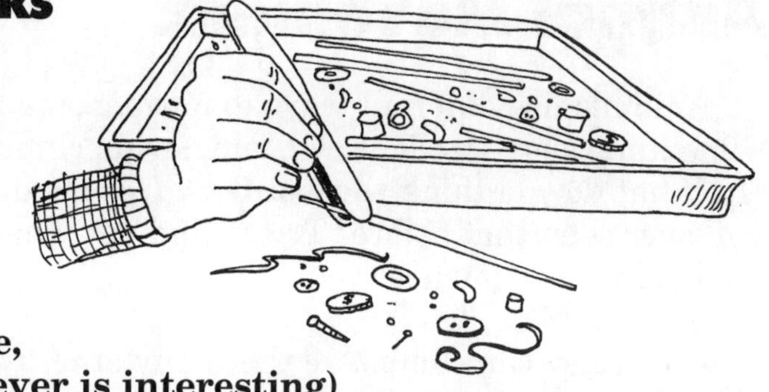

What you need:

chopsticks
ice-cream bar sticks
different sized small objects:
(spaghetti or rice—cooked and uncooked—seeds, a cookie, gelatin, marshmallows—whatever is interesting)

What to do:

1. First, use the chopsticks to try to pick up the objects. Describe what you could pick up easily. What was more difficult to pick up? What could you not pick up with the chopsticks?

2. Now use your ice-cream bar sticks like a bird's beak. Describe what was easier to pick up with those sticks than with the chopsticks. Was there anything you could not pick up?

3. Do you want to eat the cookie? Remember, you're a bird now and you can use only your beak. What will you do? Write what you did to eat your cookie. Write how using ice-cream bar sticks and chopsticks demonstrated how birds use their beaks. How did it demonstrate how larger and smaller beaks are the same? How are they different?

What are some other animals that mainly use their mouths to get food?

Here's a hint: How many fish do you know that have hands and feet? When was the last time you saw a giraffe sit down to a table with a fork and knife? What other animals can you think of that mainly use their mouths for getting food?

A Jungle Jingle

Here's a fun poem you can use to play, imitate, and guess animal movements.

Jungle Jingle

A big, gray elephant lives in the jungle,

And when he walks he walks with a rumble.

He tosses his trunk high in the sky,

Kicks his back legs, and winks one great eye,

Then slowly goes back to the jungle.

Animal Movement

Animals use different ways to move. Some hop, some jump, some walk, crawl, run, swim, climb and/or fly. What do you think the words *and/or* mean? Have you ever seen that before? Try to find out what and/or means in this sentence.

A monkey can climb. Are there any other ways you think a monkey can move? A duck can fly, but what other ways can a duck move? What about you? How can you move?

Imitating Animals

What you need:

a partner or two to guess your pantomimes

What to do:

1. Pretend to be the elephant in the jingle on page 25.
2. Write down your own jingle for another animal.
3. Ask your partner to write a jingle for an animal, too.
4. Pantomime your jingle and try to guess each other's animal.

Skeletal and Muscular Systems

In most animals, their skeletal and muscular systems help them move. Some animals such as jellyfish, worms, and octopuses don't have skeletons at all. In other animals, such as lobsters, insects, and spiders, the skeleton is on the outside of their bodies. In animals such as birds, reptiles, amphibians, fish, and mammals, the skeleton is on the inside. Write what you think the skeletal system is. Write what you already know about the muscular system.

____

____

____

Which animals below have no skeleton at all? ____

____

Which animals have a skeleton on the outside, called an exoskeleton? ____

____

Which animals have a skeleton on the inside? ____

____

____

Portuguese Man-of-War (Jellyfish)

bird

fish

spider

insect

earthworm

crab

bear

alligator

How Many Bones and Muscles Do You Have?

Bones and muscles work together to help animals. Together they help protect the body and help it move. Can you believe you have 206 bones in your body? You have 604 body-moving muscles that are attached to those 206 bones. You have other muscles, like your heart, that are not attached to bones, too.

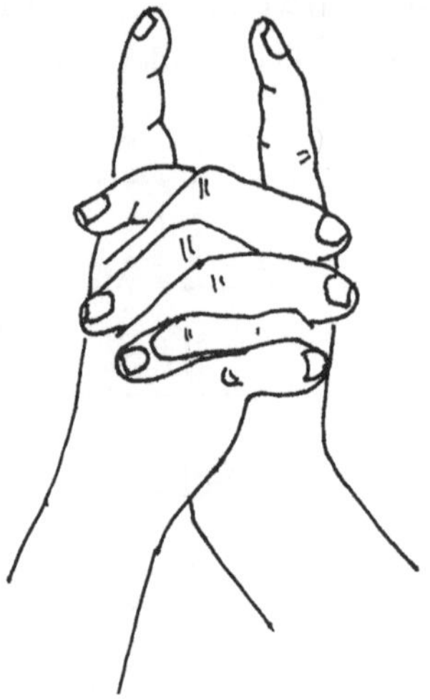

Muscles Relax, Too

Your heart is working all of the time. But other muscles relax sometimes. Test the positions in which your muscles relax.

What you need:

your fingers

What to do:

1. Clasp your hands together. Interlock your fingers.
2. Hold your index fingers straight up. Relax. Are your fingers perfectly straight? Do you have to work to make your fingers perfectly straight?

Muscles and Bones Are Spectacular Machines

Did you know that many of your bones work like machines? Do you remember what a lever is? It is a bar, rod, or plank that pivots on one place called a fulcrum. Many of your bones work like levers when the muscles move them. They lift and move different parts of your body the way levers lift and move objects. The lever moves a load using some type of force. That's a pretty confusing explanation, isn't it?

Let's see if we can make it a little easier to understand. Have you ever played on a seesaw? That is one type of lever. When you lift an adult much heavier than you, you are the force. Your adult partner is the load. The part you sit on is the lever. And the part holding up and balancing the seesaw is the fulcrum.

Look at the picture to see the different parts of the seesaw lever.

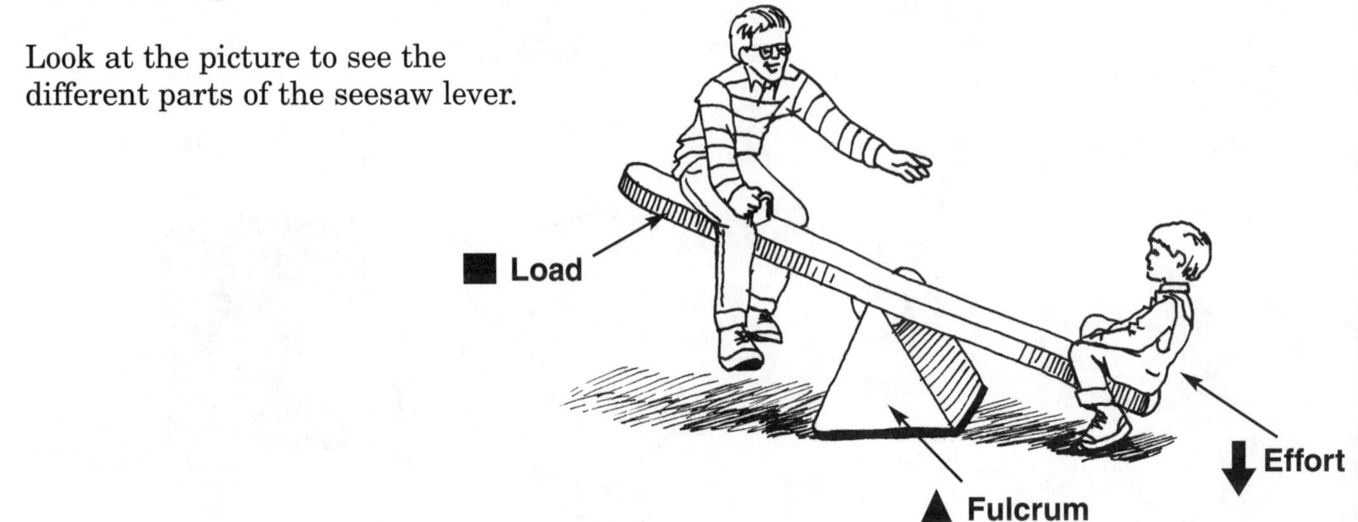

Your Head and Neck at Work

Your head and neck joint are an example of this kind of lever in your body. Your neck muscles nod your head up and down on the neck joint.

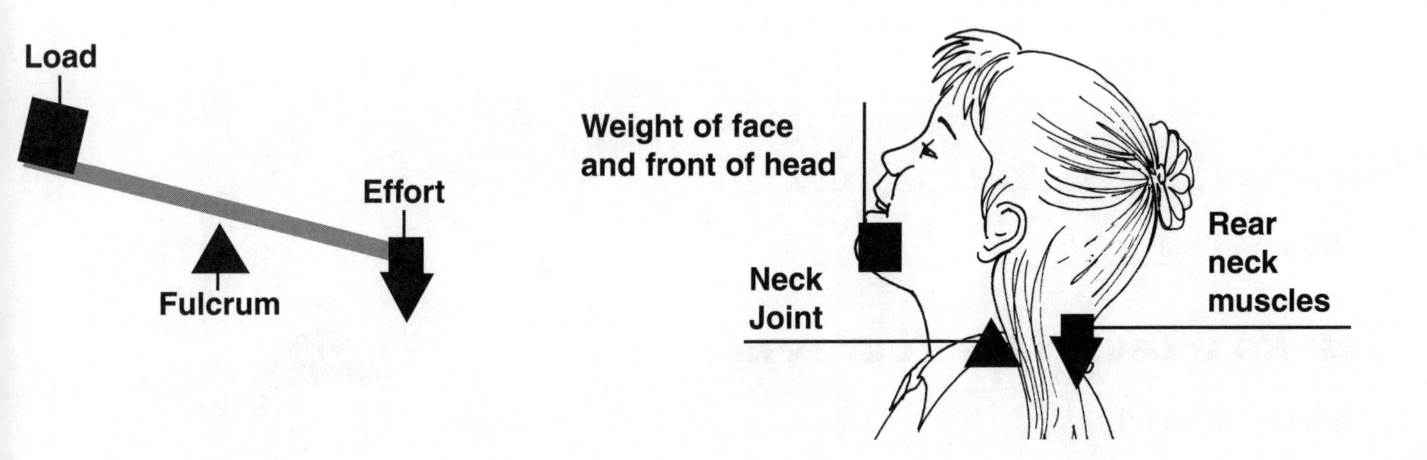

You Can Make Your Muscles Work

You can make your muscles continue to work, even after you stop working. Let's see how.

What you need:

a door frame **your muscles**

What to do:

1. Stand next to the door frame and press very hard against it with the back of your hand. Look at the picture.
2. After 30 to 60 seconds, step away from the door. What happens to your arm? Why do you think this happened?

Your Elbow, the Fulcrum

Your elbow, lower arm, and hand are a different kind of lever. Lift your book. When you lift your book, the book is the load. The muscles in your arm are the force. And your elbow is the fulcrum. Referring back to page 29, write what the book is compared with in the seesaw lever. What are the muscles in the arm compared with in the seesaw lever? What is the elbow compared with in the seesaw lever?

book = ______________________________

muscles = ______________________________

elbow = ______________________________

Make a Lever Like Your Arm

What you need:

poster board
ruler
modeling clay
a pin
2 small metal screw eyes or brackets
scissors
pencil
string or yarn

What to do:

1. Cut out two strips of poster board, 2 inches by 10 inches. Draw a line down the center of each strip. Start at one end of one strip and make a mark 6 inches on the line. Start at one end of the other strip and make a mark at 1 inch.

2. Screw one screw eye into each of the two marks you made on the two strips. At the other end of each strip make a mark 1 inch from that end. This will be the elbow joint.

3. Put the new pencil marks on top of each other. Make sure the screw eyes are on the same sides. Push the pin through the new pencil marks and put the modeling clay on the sharp end so you don't stick yourself.

4. Tie the string or yarn to the screw eyes. Pull on the string to make the arm move.

The string is like your arm muscle. The strips of poster board are like your upper and lower arm skeleton. The pin area is like where your arms join together at the elbow. There are more muscles in your real arm that make it move. Remember, this is just a simulation. Try to add more screw eyes and strings to your model to simulate the muscle in your lower arm.

Thigh Bone Connected to the Leg Bone

You can look at how muscles and bones work together in a real chicken leg.

What you need:

a whole chicken leg (leg and thigh still attached) **an adult partner**

What to do:

1. Ask your helper to remove the skin from the chicken leg and thigh.
2. Pull one muscle and see what happens to the bones and to the other muscles. Describe in writing what happens.
3. Find the joint between the thigh and the leg. Look at how it is joined. Describe in writing and draw the joint, the muscle, and the bones.
4. Find cartilage covering the ends of the bones. Describe what the cartilage looks and feels like. (Remember to wash your hands with soap and water after handling the chicken.)

Describe in writing in what directions the bones can move easily. Draw a picture. Describe how the muscles moved the leg. Draw a picture. Try to infer what cartilage does. Do you remember what the word *infer* means?

Cartilage Reduces Friction

Move your arm and feel the elbow joint. It moves pretty smoothly, doesn't it? Something else in your body helps you move. In most joints, the ends of the bones are protected by smooth, soft cartilage and an oily fluid. Together they reduce friction between the bones. Do you remember what friction is?

What you need:

a large baking tray
cooking oil
scissors
block of wood (unfinished)
plastic wrap
tape
a wooden board (unfinished)

What to do:

1. Lay the wooden board in the baking tray. Press the block of wood hard against the wooden board and slide it. Describe what it felt like. This is sort of like two bones rubbing against each other without any cartilage or fluid.
2. Wrap the board in plastic wrap and tape it so it won't move. Wrap the block in plastic wrap and tape it so it won't move. The plastic will act like cartilage does in a body joint.
3. Smear cooking oil over the plastic wrap on the board. Now slide the block over the oil and plastic wrap. Describe how this feels compared with when you slid the pieces of bare wood across each other.

There is still friction between the block and the wooden board. But the friction was decreased a lot by the plastic wrap and the oil. In the body, the cartilage and the fluids between joints reduce the friction between the bones.

The Jaw, a Powerful Joint

Your jaw is another kind of joint. Let's see how it works.

What you need:

1 cracker
2 stickers to put on your face
a desk top or tabletop
a helper

What to do:

1. Put one sticker on your chin. Put the other one on your cheek.
2. Break the cracker in half. Rest your nose on the table and start chewing half the cracker slowly.
3. Ask your helper to watch the stickers. Describe how they move. Describe how it felt trying to chew with your nose on the table.
4. Put your chin on the table. Slowly, chew the other half of the cracker with your chin on the table. Ask your partner to watch the stickers again. Describe how they move. Describe how it felt trying to chew with your chin on the table. Which was easier?

Compare your jaw with a lever. What part of your jaw do you think is the fulcrum? What do you think is the force? What do you think is the load? Do you remember the three different kinds of levers? One is like a seesaw. One is like a nutcracker. One is like tweezers. Which of these is your jaw most like?

Describe in writing and with an illustration, how you think your jaw works. Ask your partner to do the same experiment and observe your partner.

____

____

____

____

____

____

____

____

____

My Illustration

Where Do You Live?

Plants and animals are different in different parts of the world. Where do you live? Do you have palm trees? Do you have flying animals? Do you have pine trees, sometimes called conifers? Do you have animals that live in forests? Do you have plants that flower? Do you have animals that live in rivers, streams, and oceans? Do you have trees that flower and produce fruit? Do you have farm animals around you? Do you have any pets?

Try to answer these questions by doing some research. How do you think you could find out about your ecosystem? Now there's a great word—*ecosystem*. Do you know what ecosystems are? We'll get back to that word later on.

Four Seasons

Different places have different kinds of weather, but every place has four seasons. Name the four seasons and draw a picture to show what happens where you live during each season.

The Season Fall and Its Effect

Do the leaves turn color and fall off some of your trees during the "fall"? ______________

Where do you think the word *fall* came from? How can you find out?

____

Do you see more animals in your summer or in your winter? How are the animals you see in the summer different from the animals you see in the winter?

____

How is the weather different in your summer than in your winter?

____

List some kinds of plants and animals that live in your environment.

____

Write how the plants and animals are different and the same during the different seasons in your environment.

____

____

Different Weather

The weather is usually safe for you to be outside and observe it. But sometimes it can be dangerous to be outside in certain kinds of weather. Here are some hints about ways of observing weather and what you should do when certain kinds of weather patterns come close to where you live.

Cold

Wear warm clothes. Wear mittens or gloves and a hat. Remember, most of your body heat leaves through your head. A lot of heat leaves through your hands and feet, too. Some days in some places are too cold for you to be outside at all. Check the weather or ask an adult helper for advice. What are some ways you can check the weather for the day?

Snow

Follow the rules for cold. Try to stay warm and dry. If the snow becomes a blizzard, it will be hard to see through the falling snow. If the snow gets very heavy and windswept, a blizzard can begin and you should be inside.

Hail

Do you know what hail is? It is balls of ice produced by thunderclouds. Sometimes hailstones can get as large as golf balls, or even bigger. They can cause bruises, so go indoors when it begins to hail.

Sunny

Always remember, never look at the sun. You should use a good sun block anytime you are in the sunshine. But sunny days are good days to do a lot of nature observation.

Thunderstorm

Stay inside when there is lightning and thunder. If you are outside, stay away from tall objects and trees, and get inside as quickly as you can.

Tornado

Stay inside when you hear a tornado siren or when tornado warnings are issued. Stay away from windows, and go to the lowest floor of the house. The center of the house away from outside walls is the best place to be. Stay with an adult.

Hurricane

Hurricanes and tornadoes are not the same kind of storms. But sometimes tornadoes can develop in hurricanes, too. Follow the same instructions as in a tornado. The wind in both tornadoes and hurricanes is very, very powerful.

Catching a Snowflake

If you live where it snows, here is a fun experiment to do. You've probably heard that all snowflakes have six sides and that no two of them are exactly the same. Here's a way to catch some and keep them long enough to take a look.

What you need:

a magnifying glass or hand lens
a cardboard box with a top (like a cake box or shirt box)
a glass plate
clear nail polish

What to do:

1. The next time it is supposed to snow, place the plate and the bottle of nail polish inside the box and put the cover on the box. Leave the apparatus outside where it will stay cold.

2. When it begins to snow and the plate is very cold, pour a bit of the nail polish on the plate so a thin pool forms.

3. Wear gloves so your hands don't warm the plate, and put the plate somewhere where it will catch snowflakes.

4. When snowflakes have fallen on the nail polish, put the top back on the box and let your apparatus stand outside until the polish is dry.

5. When the polish is dry, bring the box inside and take a good look at your snowflakes with a magnifying glass or hand lens. Now you can draw what real snowflakes look like and make cards of your drawings to send to friends.

Snowflakes and Sleet

Even if you don't live where it snows, you can do some experimenting with how snow falls. Some snowflakes seem larger than others. Large snowflakes sometimes actually fall more gently and more slowly than sleet that weighs the same but seems smaller. Sleet is frozen rain. Here's how you can test that with construction paper.

Which Falls Faster?

What you need:

a sheet of construction paper **scissors**

What to do:

1. Cut your paper into about 30 squares exactly the same size. Separate your squares into two piles.
2. Leave one pile of squares as they are. These are your snowflakes. Crumple the squares in the other pile into little balls. These are your sleet.
3. Stand up as tall as you can with some flat snowflakes in one hand and the crumpled sleet in the other. Now drop your snowflakes and sleet at the same time. Record which hits the floor first.

Do your imitation snowflakes and your imitation sleet weigh the same? How do you know? What is different between the two? Now, can you guess why one hit the floor before the other?

____

____

There is more surface space on the imitation snowflake, so the air resistance is greater. Think of a parachute. How do you think a parachute keeps a person from falling too quickly?

____

____

____

____

____

____

____

Air

Wind has a lot to do with weather patterns. Wind can do some real interesting things. Even though you can't actually see air, it is very powerful. You may think you can see air, but what you really see are dust, dirt, leaves, and other substances moving in the air. Here's a way to see the power of air and the pressure air can exert.

You can surprise a friend with this experiment. It is almost like a magic trick.

Equalizing Air Pressure

What you need:

a ruler
4 sheets of newspaper

What to do:

1. Lay the sheets of newspaper on a flat table top. Slide the ruler under the sheets as shown. Leave less than half of the ruler sticking out over the edge of the table.

2. Ask your friend to lift the paper by smacking quickly down on the ruler with his or her open hand. (Tell your friend to slap the ruler quickly but not too hard because the ruler might snap in half.)

3. Try it. Now you move your hand slowly down and pretend to slap but just slowly and gently put pressure on the ruler. Now the paper should lift up.

Here's what happens. To lift the paper, you need to move a lot of air out of the way. Doing that quickly takes a lot of force. When you do it slowly, you can do it with one finger. When you press down slowly on the ruler, the air has time to move in under the newspaper and equalize the air pressure under the papers with the air pressure above the papers.

Does Air Have Weight?

You see substances in air, substances like dust and dirt. But you can't really see the air. But do you think that air has weight? If you hold your hand out and lift in the air can you feel any weight? Is the air hard to lift? Unless the wind is moving, you can't even feel it. But air does have weight. Let's see if we can prove that with a scale you can make yourself.

Which Balloon Is Heavier?

What you need:

two balloons the same size	**a table**
a yardstick or meterstick	**string**
a pencil	**tape**
several heavy books	**a pin**

What to do:

1. Blow up both the balloons to the same size. Tape one balloon to each end of the yardstick (meterstick).

2. Tie a string to the exact middle of the yardstick (meterstick). Tie the other end of the string to one end of a pencil.

3. Place the other end of the pencil under books on a table so the string and yardstick (meterstick) are hanging from the end of the pencil.

4. Adjust the yardstick (meterstick) until the balloons are balanced. If you blew them up exactly the same size, they should balance easily.

5. Now pop one of the balloons with the pin and watch what happens.

Draw your apparatus and describe in writing what happened when you popped the balloon. Try to explain how this proves that the air in the balloons had weight.

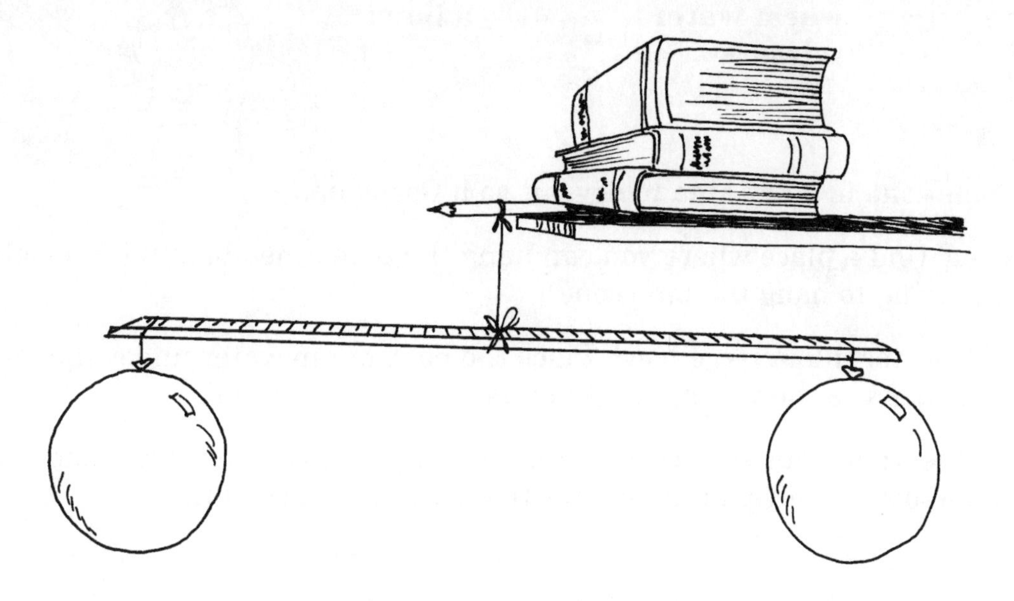

What's the Weather?

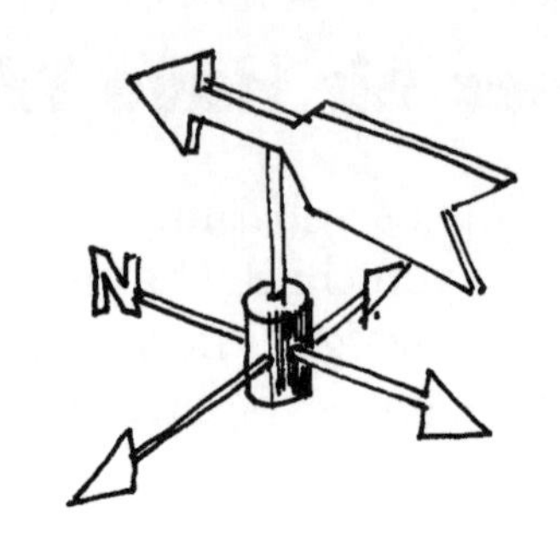

People have always tried to predict weather. People called meteorologists try to predict the weather today with some very precise instruments like hygrometers, barometers, anemometers, and even with satellite photographs from space. Before all these inventions, people predicted weather by observing nature.

How hot does it get where you live? ______________ How cold does it get? ______________

What do you do on a hot day? Do you try to cool off? Tell how.

____

What do you do on a cold day? Do you try to warm up? Tell how.

____

Heating up and cooling down are very similar in some ways. You can put on a sweater, a coat, mittens, and a hat to keep the warmth in your body. Did you notice that sentence said "keep the warmth in your body"? It didn't say "keep the cold out of your body." When you cool off, your body feels like it is getting cooler because the heat is leaving your body and warming the cold air around you. When you warm up, heat is entering your body from a warmer outside. Here is one example of how observing nature helps predict weather. A hygrometer is an instrument that measures humidity, the amount of water in the air.

Nature's Hygrometers

A pine cone is one of nature's hygrometers. Pine trees produce their seeds in pine cones. The seeds are carried by the wind to new growing places. As the pine cone dries, it opens to let the seeds fall out and float in the wind. But when the air gets humid or moist, the pine cone closes again, even after the seeds are gone.

What you need:

a pine cone
string
warm water
scissors
a bowl

What to do:

1. Leave your pine cone in a dry area to dry out so it opens up.
2. When it is open, find a place where you can hang the pine cone about 6 inches above a table. Use the string to hang the pine cone.
3. Pour warm tap water into a large bowl. Place the bowl of tap water under the pine cone. Watch what happens as the warm, moist air rises to the pine cone.

Write what you observe in your science journal. Draw a picture of your pine cone when it was dry. Draw a picture of your pine cone after the moist air reached it.

Can Animals Predict Weather?

Some people still believe that animals sense changes in weather. Watch the animals where you live, and decide for yourself if any of the ways people believe animals behave predicts rain.

Here are some signs that are supposed to mean rain is on the way. Do you believe them?

Frogs croak louder and longer than usual.

Dogs whine or act nervous, and cats get frisky as kittens.

Roosters crow later in the day.

Birds fly lower to the ground and gather on tree branches and telephone wires.

Pigs squeal more and gather sticks to make a nest.

Cows sit down in the fields to feed. They run around the field with their tails high swatting flies before a storm.

Bees and butterflies seem to disappear from the flowerbeds they usually visit.

Red and black ants build up the mounds around their ant holes.

Fish jump out of the water to nip at low-flying insects.

Some crickets are said to chirp only when rain is on the way.

Fat and Warmth

Animals don't put on mittens and hats when it gets cold outside. Instead, some animals have heavy fur coats and some animals build up more fat in their bodies for the cold winter. Some seals swim in very cold water. So do some whales and dolphins. They stay warm because of the fat layers they build up in their bodies. That seems kind of weird that fat could keep warmth in a seal, doesn't it? Well let's try an experiment with lard gloves. In a lot of ways, the lard will act like fat layers in seals and whales and other animals that need to keep warmth in their bodies.

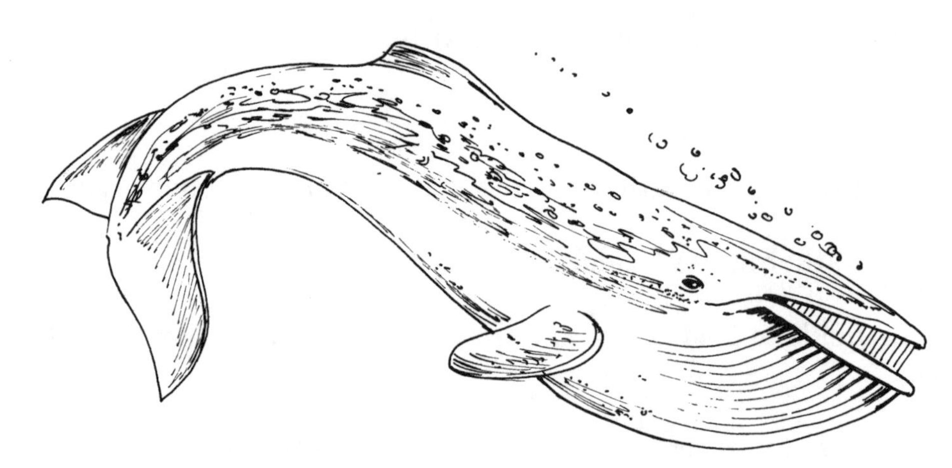

Fat Is a Great Insulator

What you need:

two plastic sandwich bags
lard or shortening
masking or duct tape
a large bowl
ice water

What to do:

1. Fill a large bowl with ice water.
2. Fill one plastic sandwich bag with lard or shortening. Put the other plastic sandwich bag on one hand.
3. Slide the hand with the plastic bag on it into the bag filled with lard or shortening.
4. Tape the bags closed around your wrist. Make sure lard or shortening is covering the layer in between the two plastic bags so your hand is surrounded by the lard or shortening.
5. Put both hands in the ice water. What difference do you feel between your two hands?

Do you think this apparatus would make good mittens in the cold? They might be a little messy. What did the lard or shortening do for your hand in the ice water? What do you think a layer of fat does for a seal in icy water?

Can Your Hands Measure Temperature?

What you need:

paper towels
paper
pen
tape
a pan of very warm water
(like you use for a bath)
a pan of very cold water
a pan of room temperature water

What to do:

1. Mark the very warm water pan #1. Mark the very cold water pan #2. Mark the room temperature water pan #3.
2. Place your right hand in the very warm water in pan #1. Place your left hand in the very cold water in pan #2. Keep your hands in the water for about 1 minute. Observe and think about how each hand feels.

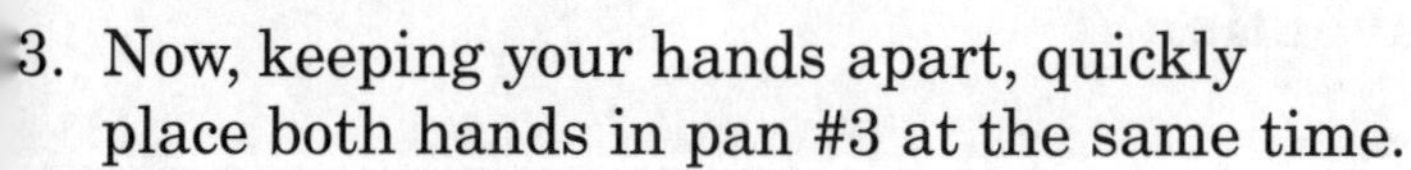

3. Now, keeping your hands apart, quickly place both hands in pan #3 at the same time.

Observe and think about how each hand feels. Dry off your hands and describe in writing how your hands felt in each pan. Describe in writing how they felt when you put both in the room temperature water. Describe why you think feeling warm and cold with your hands is or is not a good indicator of temperature.

____

____

____

____

____

____

____

____

____

Try to Invent Something

Suppose your thermos broke and you want to take some hot chocolate on a picnic today. Design a thermos that will work to carry and keep your hot chocolate hot. List the materials you will need to build your thermos. Write "What you need" and "What to do." Draw a picture showing what your thermos looks like. Include in your design picture a cross-section drawing, showing the linings of the thermos if there are any. First, what is a cross-section drawing? How can you find out before you start your design? Use this page for your design.

What you need:

What to do:

1. ______________________________

2. ______________________________

3. ______________________________

4. ______________________________

My Invention

Temperature Determines Movement

When an object is heated, its particles move faster. When an object cools, its particles move more slowly. All gases expand when they are heated and contract when they are cooled. Let's see if we can check to make sure that statement is true.

Here's one way to see if gas expands. But you'll have to do it when the heat is on, probably in the winter. You can try the second experiment if your heat is not on.

Expanding Gas Experiment #1

What you need:

a balloon
a radiator or heating vent
masking tape

What to do:

1. Blow up the balloon until it will not take anymore air without bursting.
2. Place the balloon near a radiator. Use a piece of tape to keep it there. (Be careful not to touch the radiator itself.)
3. After a while, the balloon will grow larger and eventually pop as long as the radiator or air vent stays warm. The heat expands the air in the balloon. The warm air expands so much the balloon eventually pops.

Expanding Gas Experiment #2

If your heat is not on, this experiment will also show you how heat expands air.

What you need:

a small clear jar with a lid
(a baby food jar)
a clear-plastic bottle
(a plastic juice bottle)
modeling clay
red food coloring
water
a clear plastic straw
ice cubes in a plastic bag

What to do:

1. Wet the top of the lid of the glass jar. Turn the wet top upside down, and place it on the mouth of the jar so the water makes a seal.

2. Put your hands around the jar and hold it to warm the air inside. Watch the lid. Describe what happens. The warming air is actually doing work. What work did the warming air in the jar do?

3. Now, fill the plastic bottle halfway with colored water. Place the straw in the mouth of the bottle and down into the water. About 2 inches of the straw should be in the water.

4. Plug the mouth of the bottle around the straw with clay so no air gets into the bottle except through the straw. Blow into the water. Describe what happens.

5. Hold your hands around the part of the bottle filled with air. Describe what happens to the water level in the straw.

6. Hold the bag of ice cubes against the same part of the bottle. Watch the water level in the straw. Now describe in writing what happens. How do you think changing the temperature of the air affected the water level in the straw?

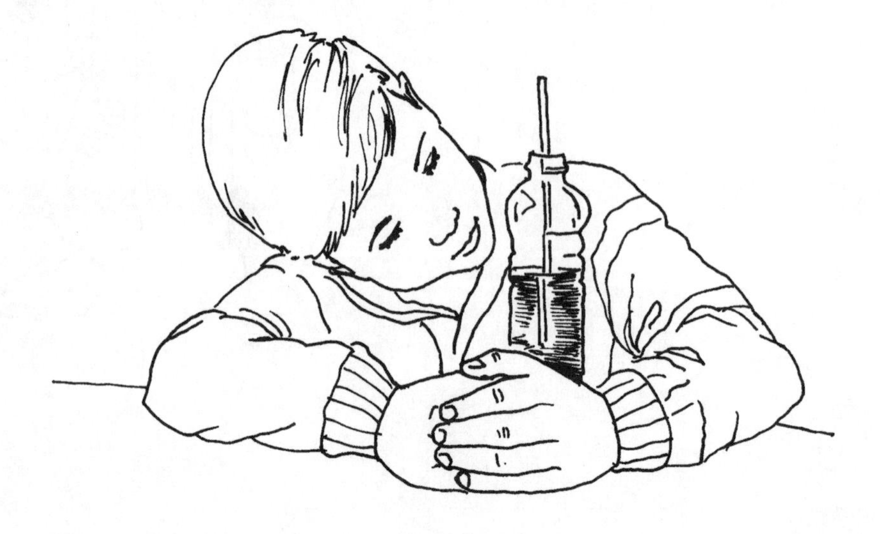

Solar Energy

The heat in the last experiments heated the air so the air did some work. What kind of work did the air do in the experiment with the jar and its lid? What kind of work did the heat do in the experiment with the colored water?

Since heat is an energy source to do work and the sun produces a lot of heat, can the sun's heat be used to do work? The heat energy produced by the sun is called solar energy. But do you think the sun's solar energy is really powerful enough to do work for us? Half the sun's energy is light and half the sun's energy is heat. Describe how the sun's energy affects you on a clear hot day. You can harness that heat to work for you. (Remember, NEVER look directly at the sun!)

The Sun's Power

What you need:

a chocolate bar
a dish
a magnifying glass
a sunny day

What to do:

1. Place the chocolate bar in the dish.
2. Place the dish with the chocolate bar in a sunny spot.
3. Using the magnifying glass, focus the sun's light on the chocolate bar.

Describe in writing what happens. Record how long it takes for some reaction to take place. What work is the sun doing in this experiment?

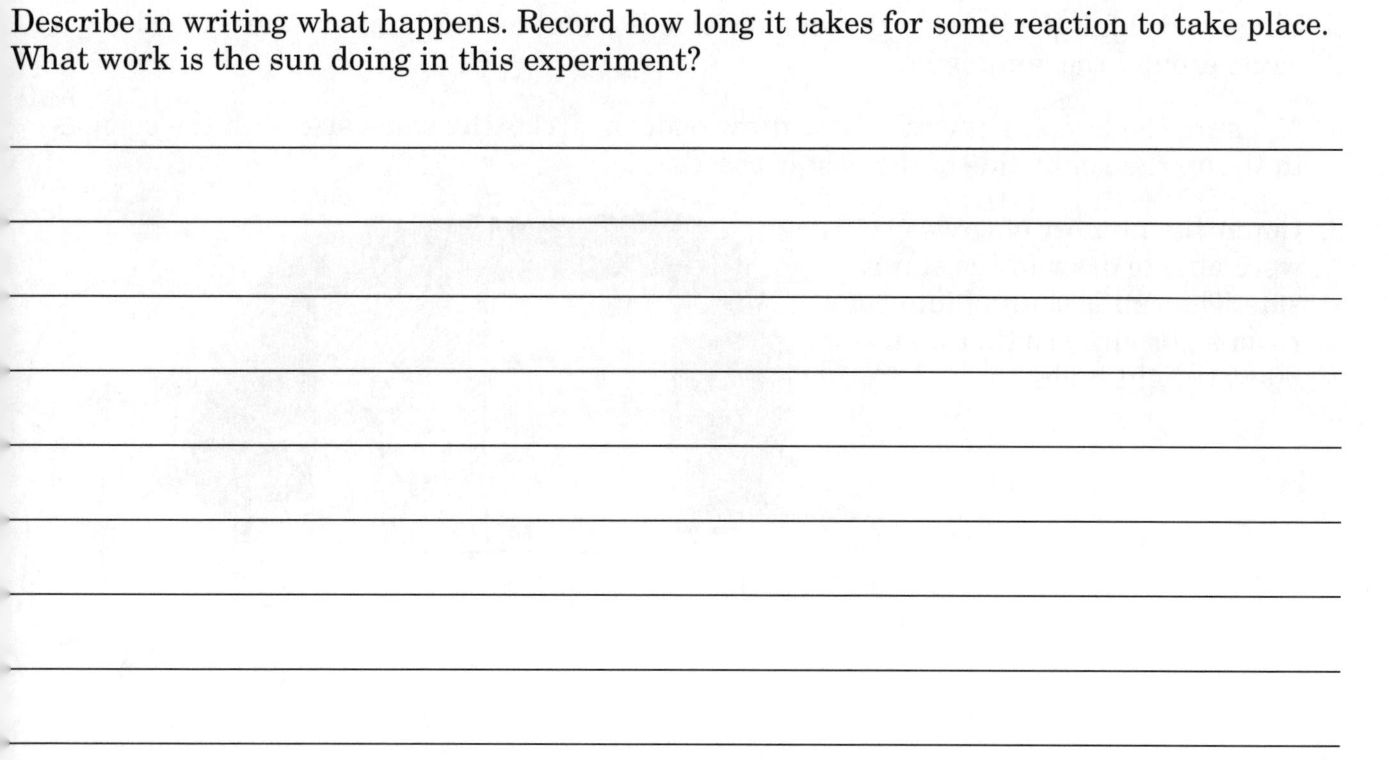

The Sun's Energy Is Extremely Powerful

Half of the sun's energy comes out as heat and half as light. We can measure its heat energy with a thermometer. But can you measure how bright the sun is? Well probably not exactly, but you can at least estimate its brightness on a given day compared with a light bulb. Let's see how you can do that.

What you need:

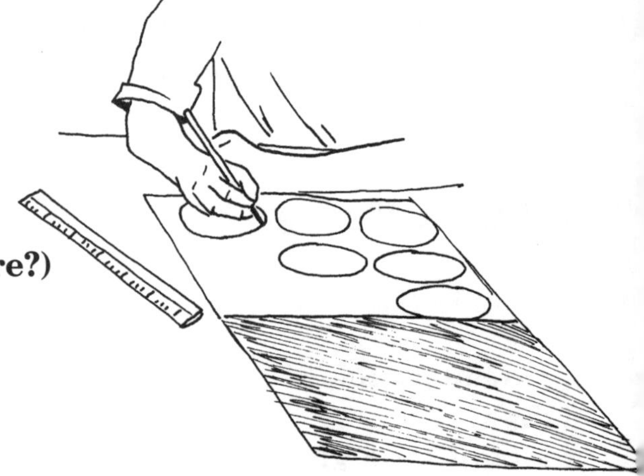

- **a lamp with a 60-watt light bulb you can point directly at paper**
- **a ruler**
- **a compass (for drawing circles) (What other kinds of compasses are there?)**
- **a pencil**
- **notebook paper**
- **white poster board**
- **a sunny day**

What to do:

1. Divide the poster board in half with a pencil line. Make sure you have two equal halves.
2. Set the poster board up against something so one half is directly facing the sun. The other side should be in a shadow.
3. Place the lamp so its light is pointing directly at the shaded half of the poster board. Adjust the lamp close enough to the poster board so the light it is reflecting is as bright as the sunlit side.
4. When you think the light on the shadow side is about as bright as the sun side, draw a circle around the lamp light.
5. Measure the lamp-light circle. Now draw enough circles the same size with the compass to fill up the sunlit side of the poster board.
6. Count the number of circles you were able to draw in the sunlit side. The sun is shining into your room as brightly as that many 60-watt light bulbs.

Simulate the Atmosphere's Effect on Sunlight

The gases in the atmosphere reflect the blue color of light from the sun. That's what makes the sky look blue. But as the sun rises or sets, the light is shining through a thicker layer of the atmosphere. The thicker layer of atmosphere can bring out colors that are more orangish or yellowish.

What you need:

a pitcher
60-watt lamp
whole milk
glass
water
a tablespoon

What to do:

1. Fill the pitcher with water. Add three tablespoons of milk to the pitcher of water. The milk will act like particles of dust and gas in the atmosphere.

2. Set up the lamp to shine into the pitcher. Now, look at the water from different angles. Look from the side. Then look from the side of the pitcher opposite the lamp. Describe how the colors are different depending on which way you look at the light shining through the pitcher.

3. Pour a glassful of the milky water. Do steps 1 and 2 with the glass of milky water. How is this different? What effect does the size of the container have on the color of the light? Try to explain how this experiment shows the effect particles in the atmosphere have on the color of sunlight.

____

____

____

____

____

____

____

____

____

____

Coax the Colors Out of a Sun Ray

Sunlight seems yellow at times. Sometimes it appears to be other colors, like orange and red. But really sunlight is filled with many colors—all the colors of the rainbow as a matter of fact. The colors in light are called a spectrum of colors. When the sun's rays and other rays of light move in and out of a transparent material like water or glass, the rays can be split into many colors of the rainbow.

What you need:

a flat, square, or rectangular pan
a mirror
a window with lots of sunlight
water

What to do:

1. Fill the pan halfway with water. Place the pan of water in a spot where the sun can shine into it through the window.

2. Put a small mirror into the pan and prop it up against one side of the pan. Arrange the mirror and pan so the sun's rays shine directly into the mirror.

3. Look for a rainbow of color on one of the walls. Each color in the sun bends in water at a different angle. When the colors bend, they spread out into a spectrum of colors.

What else have you ever seen in nature that is like the rainbow you made in your room? How do you think those rainbows might be formed? When can you usually see rainbows?

A Solar Eclipse

Once in a while the moon orbits between the sun and Earth in just the right position to block out the light coming from the sun to Earth. This is called a solar eclipse. The positions for an eclipse don't happen very often. But even if they do orbit into the right position, you should NEVER look directly at the sun—especially during an eclipse.

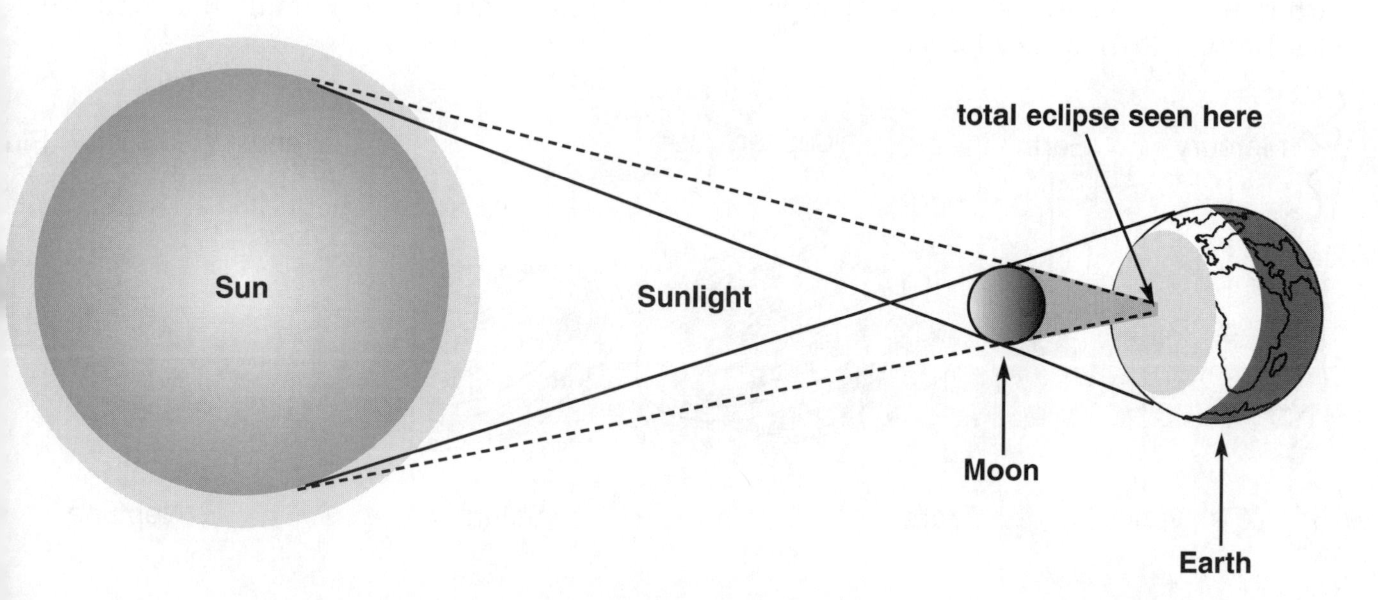

You Can Simulate a Solar Eclipse

What you need:

modeling clay
a wooden or metal skewer
poster board
a ruler
a lamp

What to do:

1. Use the clay to form a model of the moon about 1 to 2 inches in size. Place it on the skewer. Put the other end of the skewer into clay and set the apparatus on a table.
2. Set the lamp up on one side of your moon. Set the poster board on the other side. The lamp (your sun) and the clay (your moon) should be about the same height.
3. Switch on the lamp and look for the shadow on the poster board. Find the shadow's center and mark it. Mark two edges of the shadow, also.
4. Punch holes through the marks. Now, look back through the holes to see what different positions of an eclipse look like.

How do you think a small object like the moon can possibly block out a large object like the sun? Try to design a way to use your finger as a small object like the moon. Find a large object and try your design.

Our Universe

Our Sun is the center of our solar system. Do you remember the names of the planets? If not, here they are again, and here is a picture of their relative size. There's another new phrase. What does *relative size* mean? That means what something is like compared with something else. For example, compared with an elephant, you are very small. But compared with a mouse, you are very large. Compared with Jupiter, Earth is very small, but compared with Pluto, Earth is very large.

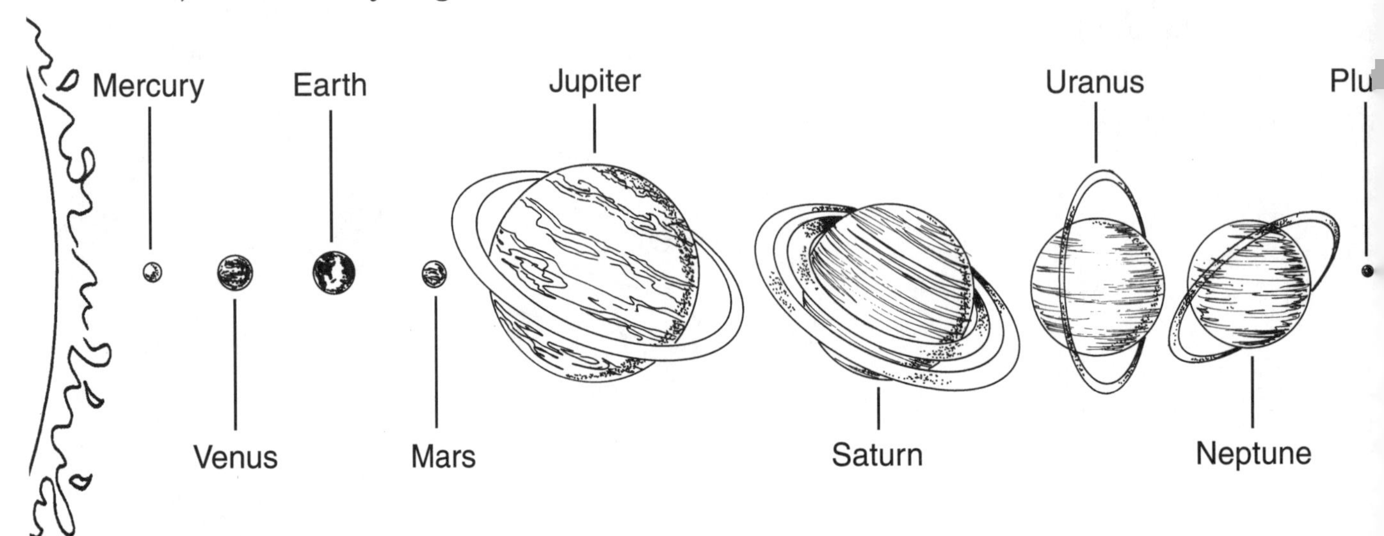

The Relative Sizes of Planets

Even seeing drawings, it is difficult to imagine the three-dimensional relative sizes of the planets. So let's try to compare their sizes with things we can touch and hold.

What you need:

½ rice grain for Pluto
peppercorns for Mercury and Mars
peas for Venus and Earth
plums for Uranus and Neptune
large orange for Saturn
grapefruit for Jupiter

What to do:

1. First, remember these are still only models, but they will give you a pretty good comparison of sizes of our planets.

2. Hold the planets in your hands and compare their sizes.

3. Put them in the correct orbital order around the Sun.

   How many Mercuries do you think it would take to make up a Saturn?
   How many Earths do you think it would take to make up a Jupiter?
   How many Mars do you think it would take to make an Earth?

All Our Planets Orbit Our Sun

The word *planet* comes from a Greek word meaning "wanderer." The planets each orbit at a different speed. How fast they orbit depends on how far they are from the sun. Mercury is the closest planet to the Sun. It orbits the Sun very fast. It orbits the Sun in three of our Earth months. But Pluto, which is the farthest planet from the Sun, takes 248 Earth years to orbit once around the Sun. You'd have to wait a long time for your birthday if you lived on Pluto, wouldn't you? You can race the planets around the sun.

What you need:

toothpicks string scissors tape
modeling clay
9 colors of pens or markers
a large sheet of poster board
colored paper that matches the pens or markers

What to do:

1. Use a string to draw a sun and nine circles around it. Use a different color to draw the nine circles around the sun.
2. Mark 12 dots on each line like the illustration shows.
3. Make 9 triangle-shaped flags, each a different color, and tape each flag to a toothpick. Stick each toothpick in a small lump of modeling clay for a stand.
4. Place each flag on a dot. Now, start orbiting. Move each flag one dot until they have all traveled 12 dots. Write which flags made more than one orbit of the sun. Which flag made one orbit? Which flags made less than one orbit?

Orbiting the Sun

Remember, this "game board" you made was only a model. The orbits of the real planets are not perfectly round. They are ellipse-shaped. Is *ellipse* a new word for you? Write it down in your science journal. It means "oval." So the planets' orbits are oval, not perfectly round.

Here's how many Earth months it takes each planet to orbit the Sun:

Mercury	3 months
Venus	7 months
Earth	12 months
Mars	23 months
Jupiter	142 months
Saturn	354 months
Uranus	1,008 months
Neptune	1,978 months
Pluto	2,976 months

Figure out how many Earth years each of these planets take to orbit the sun.

Earth takes	**1**	**Earth year to orbit the sun.**
Mars takes	____	**Earth years to orbit the sun.**
Jupiter takes	____	**Earth years to orbit the sun.**
Saturn takes	____	**Earth years to orbit the sun.**
Uranus takes	____	**Earth years to orbit the sun.**
Neptune takes	____	**Earth years to orbit the sun.**
Pluto takes	**248**	**Earth years to orbit the sun.**

The Milky Way Isn't Just a Candy Bar

There are many more suns and solar systems besides ours. As a matter of fact, the stars we can see in our galaxy are only a small number of the 200 billion stars that make up the Milky Way. That's the name for the galaxy our solar system is in. Here's another number that is almost impossible to imagine. Some scientists estimate that there are about 100,000 million galaxies in the Universe. How many zeros would be in the number one hundred thousand million?

People have always been interested in exploring and understanding our solar system and other solar systems beyond ours. And every year we are seeing more scientists get the opportunity to venture further in search of new information and exciting discoveries. Now it's time for you to explore, at least in your mind.

What you need:

your imagination
a trip to the library
a science journal
a pen or pencil (A computer and the Internet would be helpful, too.)

What to do:

1. Imagine you are a part of a mission to another planet, another solar system, or another galaxy. Decide what planet or planets in our solar system you want to explore.
2. Start finding out as much about the planet or planets as you can. How long will it take to get to the planet? What kind of atmosphere does it have? What kind of protective suits will you have to wear? What is the climate? What is the temperature? Write some questions you want to investigate.

____

____

____

____

____

____

____

____

____

____

Here's a little information about the other planets to get you started.

Mercury:	Closest to the sun. In the day +840 degrees F. At night temperature as low as –275 degrees F.
Venus:	Same size as Earth. Atmosphere's pressure is 90 times stronger than Earth's. Heat is trapped, so surface glows.
Mars:	Half the size of Earth. Two moons orbit. The "red planet." Temperature usually not above freezing.
Jupiter:	Largest planet in our solar system. Made of liquid hydrogen and helium. Sixteen moons orbit.
Saturn:	Rings surround Saturn. On Earth, rings would reach from Earth to the Moon. Eighteen moons orbit.
Uranus:	Mainly composed of water. Fifteen moons. Ten rings.
Neptune:	Huge storms. One moon is the coldest place in our solar system.
Pluto:	Smallest and farthest from the sun. One moon is known of.

Some scientists wonder if Pluto should really be called a planet. Other scientists theorize there might be another planet beyond Pluto. What do you think? How can you find out more?

3. Now, start your space exploration. When you have researched enough information, write a story about your trip to another planet, another solar system, or another galaxy.

Earth's Surface

Okay, let's come back down to Earth for a little while. Let's start with what makes up the largest part of Earth's surface—water. Water covers about 70 percent of Earth. Water is the only ordinary substance commonly found in three different forms. Its most common form is liquid such as rain, lakes, rivers, and oceans. As a gas, water is harder to see. When water evaporates, it enters the air to become vapor. The steam you see rising from a boiling pot is water vapor, tiny droplets of water. Water can also be a solid, such as ice or snow. Some solid water, such as glaciers, is created by nature.

Water has had a tremendous impact on the shape of Earth. Streams and rivers can start as raindrops or melting snow in the mountains. Or rain can soak the ground and gravity pulls the water to the bedrock. The rainwater flows along bedrock until it finds a way out as a spring. What starts as only trickles flows downhill, becoming streams. More rainwater and merging of the streams create tributaries. Eventually, most streams become rivers, and rivers fill up lakes and flow into oceans. As the water flows, it washes away rocks and soil, wearing the ground down.

Geologists estimate that it takes about 30,000 years to erode 3 feet of Earth's surface. Try to imagine how long it took for a river to erode the Grand Canyon. Actually, you can figure it out. If it takes 30,000 years to erode 3 feet, how many years do you think it took to erode the Grand Canyon, which is one mile deep? To get started, find out how many feet are in a mile.

Like streams and rivers, glaciers can also have a tremendous impact on the form and shape of Earth.

Simulate a Glacier

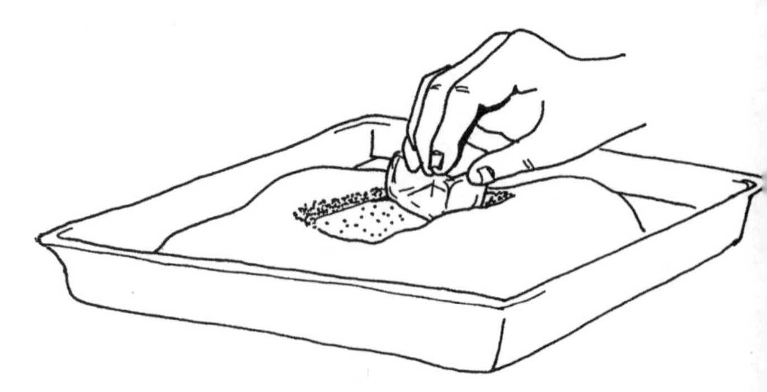

What you need:

a baking pan
sand
modeling clay
an ice cube

What to do:

1. Mold a land form in the baking pan using the modeling clay.
2. Place the ice cube on the top of the land form and push it into the clay lightly. Slide it back and forth several times.
3. Place some sand in the trench the ice cube formed. Now place the ice cube, your glacier, in the trench on the sand for a few minutes. Remove your glacier and study it. Observe and record what happened to the glacier. What happened in the trench?
4. Remove the excess sand from the clay. Rub the glacier with its embedded sand across the clay surface. Record what happens as you move your glacier. How does the sand in the glacier affect the clay?

This action is similar to what happens when rocks and other materials are dragged over land by a glacier. Now, describe how a glacier like your ice cube glacier can form a lake.

Lakes Have a Life Cycle

The water from melting and moving glaciers also forms lakes. Many lakes in North America were born more than 10,000 years ago. Water filled the basins, or depressions, left by giant glaciers. Notice the word *born*. That's kind of an interesting word to use when referring to lakes, isn't it? Actually, lakes do have life cycles similar to living things. Think about the life cycles of animals and plants: Their life begins. They grow. They age. And they die. Well, lakes also become alive through the years as plants and animals grow in and around them. The shores become homes to water fowl and other animals. The lake grows as people settle and build towns and cities. The lake provides water, food, and transportation. As time goes by, some lakes grow smaller and smaller for many reasons and eventually disappear altogether. Large, deep lakes take much longer to complete a life cycle than some smaller lakes and ponds.

Aging of a Lake

A big lake usually lives for thousands of years. But a lake ages faster if a lot of soil washes into it, or if it becomes very rich with dead plant and animal material. Let's see what different kinds of soils and growth affect the aging of a lake.

What you need:

soil and sand
2 bowls
food coloring
2 baking pans
water
sod
a watering can

What to do:

1. Fill one baking pan with soil and sand. Push a bowl down into the center of the soil and sand. Form a slope into the bowl, and make the soil and sand at the level of the bowl.
2. Place the other bowl in the center of the second baking pan. Press sod around the bowl so the sod is at the level of the bowl.
3. Use the watering can to water each apparatus thoroughly. Describe what happens in each bowl.
4. Add food coloring to the water in the watering can. Use the can to "rain" onto each apparatus. Do you see any color in the "lake"? Assume the food coloring is water pollution. Where do you think some of the food coloring (pollution) went?

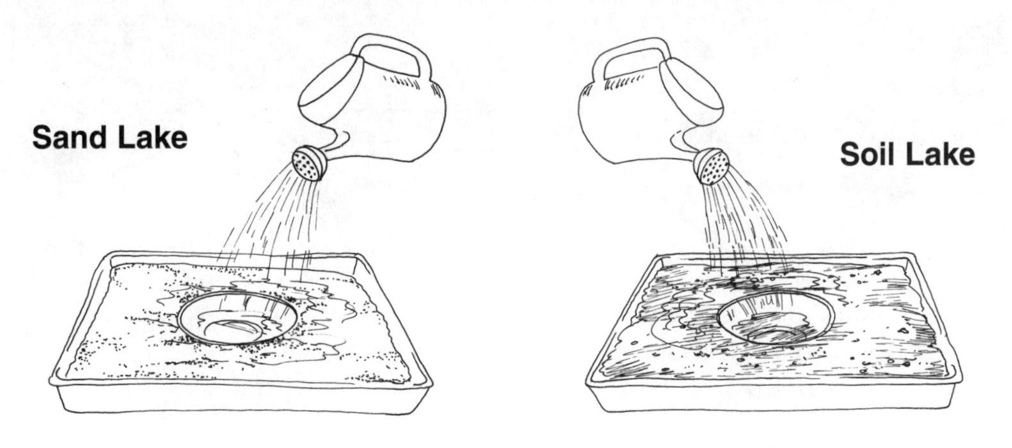

Filters, a Resourceful Tool

Filters are used to try to keep pollution and waste out of our fresh-water supplies. Filters don't work for all kinds of pollutants, such as chemical wastes that mix with water molecules. But filters do work for some solid pollutants. Let's see how a filter can work to clean up some solid pollutants.

How a Filter Cleans

What you need:

2 baking pans
water
food coloring
old handkerchief or small towel
soil
blocks, bricks, or books (Be careful not to spill water on the books.)

What to do:

1. Mix water, soil, and some food coloring in one pan.
2. Place the pan of soil and water on the blocks above the other pan.
3. Drape the cloth from the soil water down to the empty pan. Let the water from the soil pan seep into the pan below. Describe what happens. Write an explanation of why the water in the lower pan seemed clear and clean. Describe any signs of the food coloring getting into the water in the lower pan.

The water in the lower pan is still not clean enough to drink.

Acid Rain

Acid rain is caused mainly by gases released by burning in manufacturing plants. The burning produces gases that combine with water to produce an acid such as sulfuric acid. Increased acid in lakes can kill living organisms in the lake and slow down the rate at which decomposition takes place in the lake. This can cause a lake to die more quickly. A recent study showed that minerals such as sodium deep within soils neutralize acid solutions resulting from pollutants that cause acid rain. As a result, lakes fed by water that has passed through these deeper layers are less likely to become polluted by acidic pollutants. Acids in pollutants are one cause for pollution in our lakes. What are some other substances that cause water pollution? What are some ideas you have to stop the pollution of Earth's water?

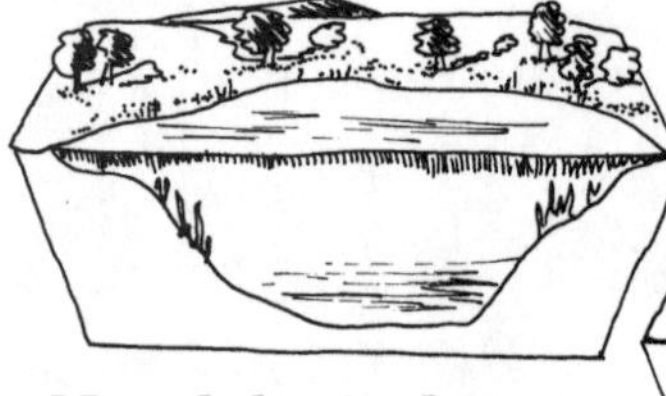

New lake is deep—acidic pollutants are neutralized.

Acids slow down plant growth and fish pollution.

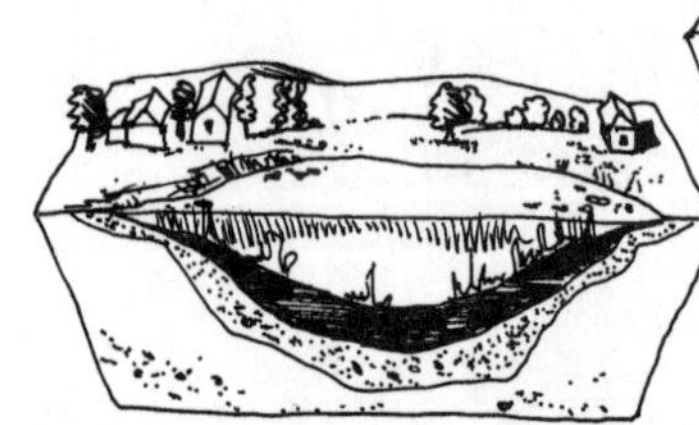

Acids slowly start to decay depth of lake and living organisms within.

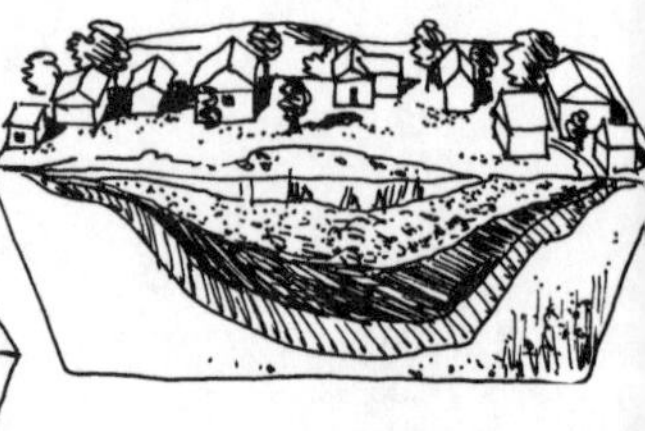

Lastly, a dead lake, which is shallow or marshy, eventually dries up.

What Do Acids Do in Nature?

Acids have many effects on different substances in nature. Limestone is one kind of mineral that can be found in rocks. Limestone is a mineral that is quickly affected by even weak acids.

Let's see what a weak acid will do to limestone. The acid we'll use is vinegar, and the limestone is chalk.

Limestone and Acid

What you need:

a piece of chalk **vinegar** **a glass**

What to do:

1. Fill a small glass halfway with vinegar.
2. Drop a piece of chalk into the vinegar.
3. Watch what happens and record your observations.

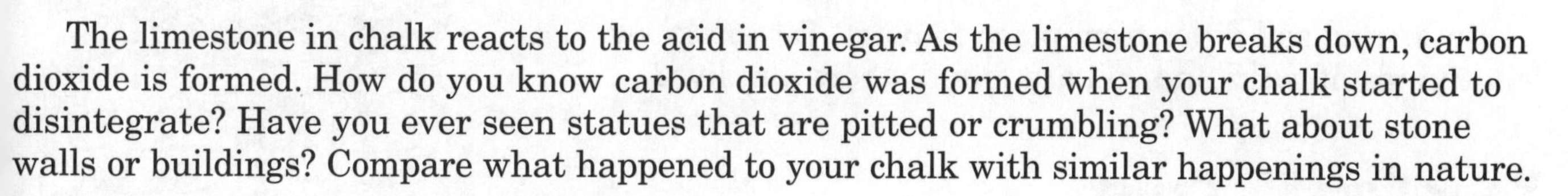

The limestone in chalk reacts to the acid in vinegar. As the limestone breaks down, carbon dioxide is formed. How do you know carbon dioxide was formed when your chalk started to disintegrate? Have you ever seen statues that are pitted or crumbling? What about stone walls or buildings? Compare what happened to your chalk with similar happenings in nature.

Calcium, an Element

Acids in pollution can affect animal life, too. Calcium is one of the most common elements found in nature. It is found in milk and in bones, as well as other places in nature. Calcium, like other elements, is affected when it reacts with an acid. Let's see what even a weak acid like vinegar can do to the calcium in an eggshell.

Making a Rubber Egg

What you need:

a hard-boiled egg **water** **vinegar** **a measuring cup**

What to do:

1. Ask an adult partner to hard boil an egg for your experiment.
2. Mix a half cup of water with a half cup of vinegar. Put your unpeeled egg in the cup and let it sit for 24 hours. Then carefully take your egg out of the solution and wash it off.

You can tell your friends that your egg came from a rubber chicken as a joke. If they don't believe it came from a rubber chicken, ask them to explain it. The acid in the vinegar dissolved the calcium in the eggshell. After a day in the solution, the thin remaining shell has been softened by the vinegar so it becomes soft and rubbery. Vinegar is a weak acid. Imagine what stronger acids in environmental pollution could do.

The Land We Live On

Earlier you learned that water covers 70 percent of Earth's surface. The rest of Earth's crust is land, land that differs in many ways from place to place. The tallest places on land are mountains. Some mountains are as high as 29,000 feet. Some mountains are higher than others. Find out the name of the highest mountain in the world. What country is it in? What mountain is the closest to where you live? How high is it?

Some high places on Earth have flat tops. They are called plateaus. What part of the word *plateaus* gives away the flatness of the area? Denver, Colorado, is built on a large plateau.

Earth's crust also has flatlands called plains. A lot of the plains on Earth are used as farmland. There is a large flatland in the middle of the United States called the Great Plains.

Some places on Earth's crust are lower than the land around them. These are called valleys. The Ohio River Valley is one example.

Look at a map and try to find the plateau where Denver is located. Find the Great Plains and the Ohio River Valley, too.

Rocks and soil differ from place to place. Do you know what kinds of rocks and soil are found where you live?

Your Earth's Crust

Let's find out what some of the materials are in Earth's crust where you live.

What you need:

a small plastic bag
spoon or garden trowel
piece of coarse screening
white paper
paper towels
a pencil
hand lens

What to do:

1. Fill the plastic bag with dirt, soil, and rocks from different locations around where you live.
2. Use the screen to sift through the materials. Use the spoon to push the small pieces of Earth's crust through the screen onto the white paper.
3. Place the pieces of Earth's crust that did not sift through the screen on another sheet of white paper.
4. Study what you collected. Use the hand lens to look closely at the materials you collected.

Your Soil Collection

What do the soil and rock look like?
How do they feel?
Are they moist or dry?
Squeeze some of the softer materials in a paper towel and describe what you see. Find some other ways to test your soil collection.
Is your soil sandy?
Is it light and does water drain through it easily?
Is your soil dark in color?
Does it retain water well?
Is your soil heavy and claylike?
Is it kind of sticky like clay when wet?
Is your soil chalky, thin, and stony?
Does water drain quickly through it?

Devise a plan to find out what type of soil you have where you live. Then find out.

____

____

____

Ecosystems

What is an ecosystem? All the living things in a region and all the soil, water, and other nonliving things they use make up an ecosystem.

Climbing a tall mountain is like passing through almost all the main ecosystems. You begin your journey in forests on the lowest parts.

You climb through grasslands, tundra, and finally snow at the top. But there are other ecosystems, such as seashores, oceans, rivers and lakes, wetlands, and deserts, too. Even towns and cities are considered ecosystems.

Look at the word *biosphere*. What parts of that word do you know? Do you know what a sphere is? Do you know what *bio* means?

____

____

Earth's biosphere is all the area of the Earth where life can exist. This includes the layer of breathable air called the atmosphere. To study Earth's biosphere, scientists called ecologists classify places on Earth into smaller areas called ecosystems.

Polar and Tundra Systems

Polar and Tundra lands are at the top and the bottom of Earth. They surround the North and South poles and are the harshest ecosystems on Earth. Temperatures can be as low as –112 degrees F and winds can blow 200 miles per hour.

The North Pole is called the Arctic. The South Pole is called the Antarctic. The coldest and harshest areas are the polar regions. The tundra area is on the edges of the frozen polar region. The tundra surface thaws out only in the summer when lichens, moss, and small bushes can grow for a short time.

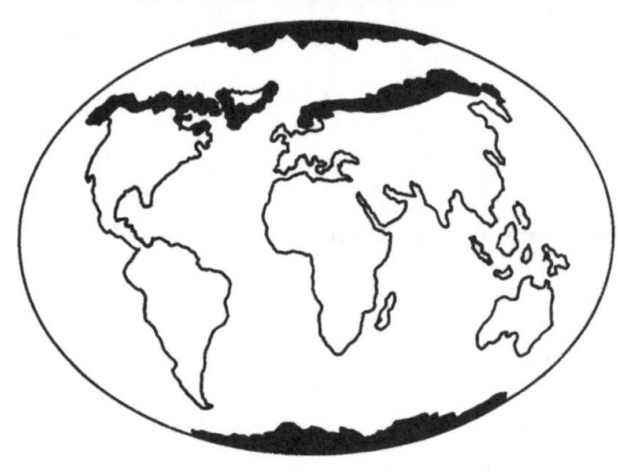

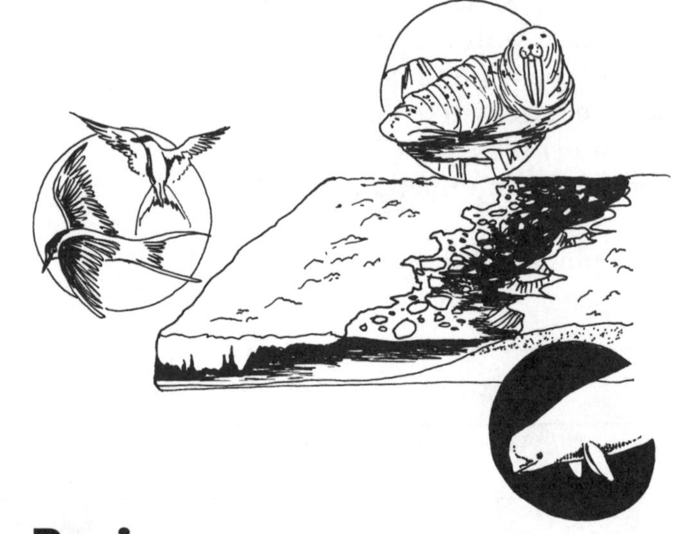

Pollution Can Melt the Polar Region

Do you remember the "Greenhouse Effect"? If Earth warms enough, the polar regions can melt. If they do melt, the melted water could raise oceans enough to cover many land areas.

What you need:

baking tray or glass dish
water
ice
modeling clay

What to do:

1. Build a small island like the one in the picture. Build a larger continent at the edge of the dish.
2. Fill the dish halfway with water to form the sea. Make sure the island and the continent surfaces are above the seawater.
3. Measure the depth of the "sea." Put ice cubes on top of the continent to simulate the ice over the Antarctic or Arctic. Then watch what happens.

How do you think this model simulates what global warming could do to the polar regions and to islands and continents?

Ocean and Seashore Systems

Oceans make up nearly 70 percent of Earth's surface, but oceans do not have a great many species of plants or animals relative to the amount of surface it covers. Only about 20 percent of known species live in the oceans even though oceans cover 70 percent of Earth's crust.

Living things exist both in the water of the oceans and on the ocean's floor. The "light zone" reaches anywhere from a few feet in muddy water to as deep as about 340 feet. Some parts of the ocean are deeper than 19,000 feet.

Seashores are where the land meets the sea. Here both the sea and the land are rich with food. The environment at seashores is constantly changing because the tides come in and go out every day. Sometimes during the day or night the tide is high, covering the seashore. Sometimes the tide is low, exposing plants and animals to air, wind, and rain.

■ **Oceans and Seas**

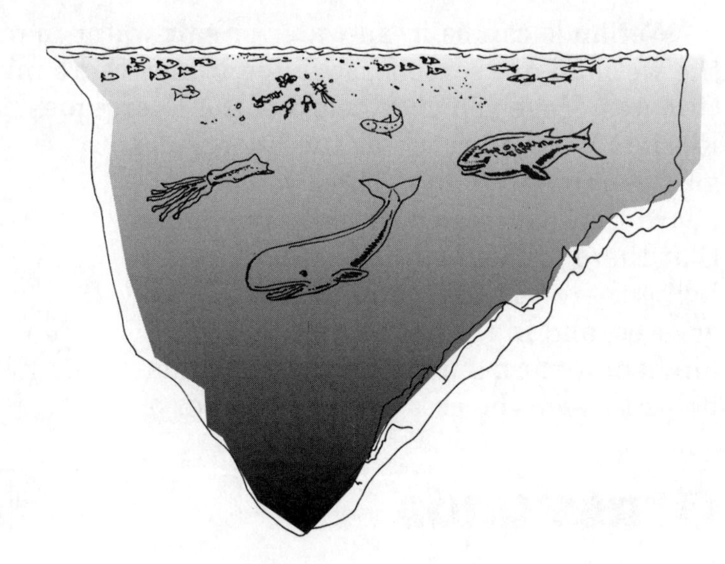

Ocean Currents

The winds over the oceans are pretty predictable. This is because there is no land mass such as mountains to change the direction. The winds over the oceans have some effect on waves, surface currents. The wind makes the surface currents move in the same direction as the wind.

What you need:

toy boat	**cork**
water	**large pan or bowl**

What to do:

1. Fill the container with water. If you want, add a little food coloring to make it look like a clear blue sea. Let the water surface settle until it is totally calm.
2. Launch the boat and the cork on one side of the pan or bowl.
3. Blow gently on the surface of the water, not on the boat or cork. Watch what happens to the boat and cork as the waves start to move.

How do you think this simulation is like the effect wind has on the surface currents? What are some things that are different?

Rivers and Lakes

Rivers and lakes form fresh-water ecosystems with many varieties of plants and animals. Some pools and rivers appear only in rainy seasons, so simple communities of plants and animals live there. But in large rivers and lakes formed over hundreds of years, complex communities exist in balance.

River and lake ecosystems are found almost everywhere on Earth. What are some rivers and lakes you know of? Are there any rivers where you live? Are there any lakes where you live? How do you think you can find out about a river or lake near where you live?

Wetlands

Wetlands can be fresh-water or salt-water marshes. Here the water levels change with the wet and dry seasons, so living things there must learn to survive in both wet and dry climates. Have you ever heard of the Everglades? The Everglades are wetlands in southern Florida. Wetlands such as the Everglades are "teeming with life." Have you ever heard that expression before? It means that there are lots and lots of lifeforms living there—in the water, in the tall grasses, and in the ground. So many kinds of lifeforms living in the wetlands helps to keep the ecosystem balanced.

Grasslands

Grasslands are dry, but not nearly as dry as deserts. Here there is enough rain for grasses and low shrubs to grow but not enough for most trees. In some ways, grass is the start of the food chain.

Grass survives animals' nibbling it because it grows from the bottom instead of from the tips like other plants. Grasses regenerate quickly after fires, and there are many fires in grassland areas.

Deserts

Deserts are harsh ecosystems. They are the driest places on Earth. Only a few animals and plants can survive the heat and dryness. There is not much decomposing material to make the soil rich. Due to the lack of water, animals and plants have adapted ways of storing and making the most efficient use of the water available.

Cacti store water in their wide stems. The gila monster, a reptile, stores water in fat cells in its fat tail. A camel can drink more than 20 gallons of water at one time. Then it can go for weeks without needing more water.

During the day, a desert can be more than 125 degrees F. But at night, the desert can be cool. Night is when many desert animals that burrow underground from the heat in the day come out to hunt.

Winds are harsh on deserts due to the lack of trees or other structures to slow them. As a result of winds, the sands move a great deal.

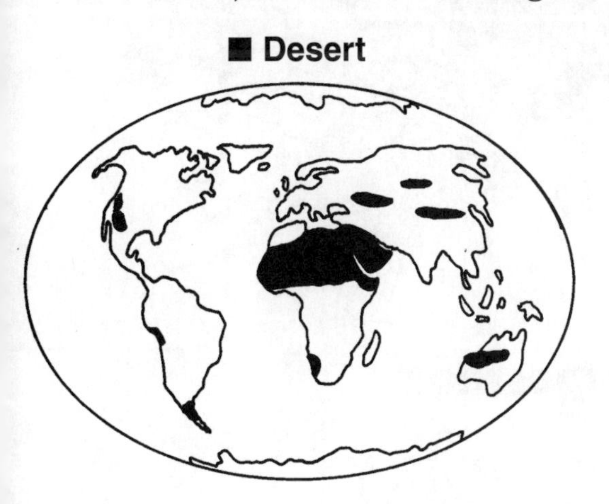

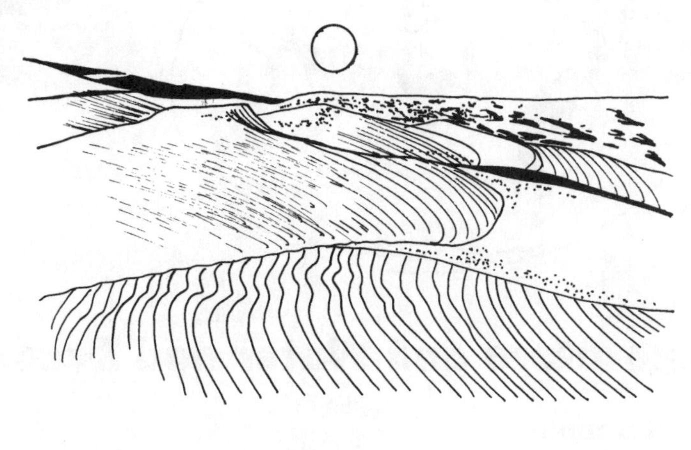

Wind and the Desert

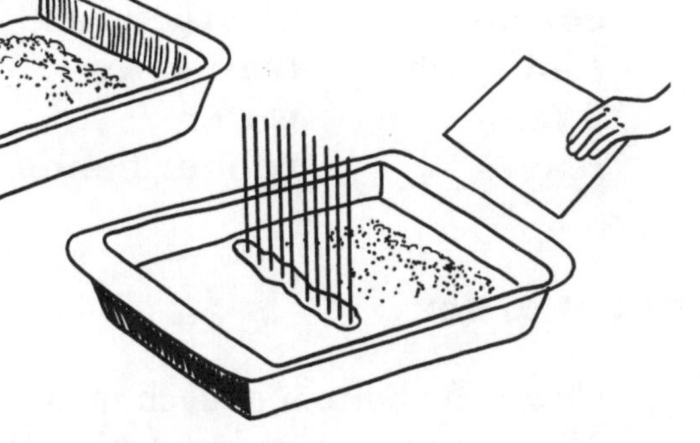

What you need:

modeling clay
a large baking tray
cardboard or poster board to produce wind
thick spaghetti
sand (dry)

What to do:

1. Place a pile of sand on one end of the baking tray. Use the cardboard or poster board like a fan to produce wind. Move the sand with wind made by the board. Record how far the sand moved without a barrier.

2. Return the sand to a pile similar to your first pile. Form a line of clay across the pan in front of the sand pile. Stand spaghetti up in the clay to form a wall like a wall of trees. Use the board to try to move the sand past the spaghetti wall. Record how far the sand moved this time.

How does this simulation show how sand can be contained? Have you ever been at the beach? If so, what kinds of structures can be used to keep the sand on the beach from eroding? What could you design to keep sand from eroding?

Tropical Rain Forests

Tropical rain forests cover less than 10 percent of Earth's surface but contain more than half of the living species on Earth. Tropical rain forests are rich in life because they are warm and wet. Rain falls almost every day in tropical rain forests, as much as 160 inches a year in some.

Thousands of species live in tropical rain forests. That's the reason so many ecologists are using tropical rain forests as their laboratories. The climates in tropical rain forests are perfect for many, many plants and animals to live. Dead plants and animals decay quickly, so the nutrients are absorbed by living plants very quickly.

Temperature and Water and Decomposition

What you need:

cotton	**plastic wrap**
toothpick	**fruit skins**
2 jars	**water**
leaves	**rubber bands**

What to do:

1. Cover the bottom of each jar with cotton balls. Moisten the cotton balls with water.
2. Put the same amount of some leaves and fruit skins on the cotton balls.
3. Place plastic wrap over each jar, and attach with rubber bands. In one jar, poke some holes in the plastic wrap with the toothpick. Put the jar with the holes in the plastic wrap in a warm place. Place the other jar without the holes in the plastic wrap in a cold place, like the refrigerator.

Record what happens to the leaves and fruit in each jar. Which simulates a tropical rain forest environment?

Temperate Forests

Temperate forests are another kind of forest. Temperate forests have broad-leaved trees such as oaks and maples and/or conifers such as pines. Most temperate forests exist where there are warm summers and cold winters. More light reaches the floor of the forest than tropical rain forests, so small plants can survive without having to creep up trees as vines.

Temperate forests can be both broadleaf trees and/or conifer trees. Conifer forests are usually found at higher elevations, such as on mountains. Broadleaf trees are also called deciduous. Usually the dry season is shorter where trees such as oaks, elms, and ash grow. The leaves fall off just as the dry season begins and new leaves grow as spring starts. The leaves, which are bigger than needles, can make food faster.

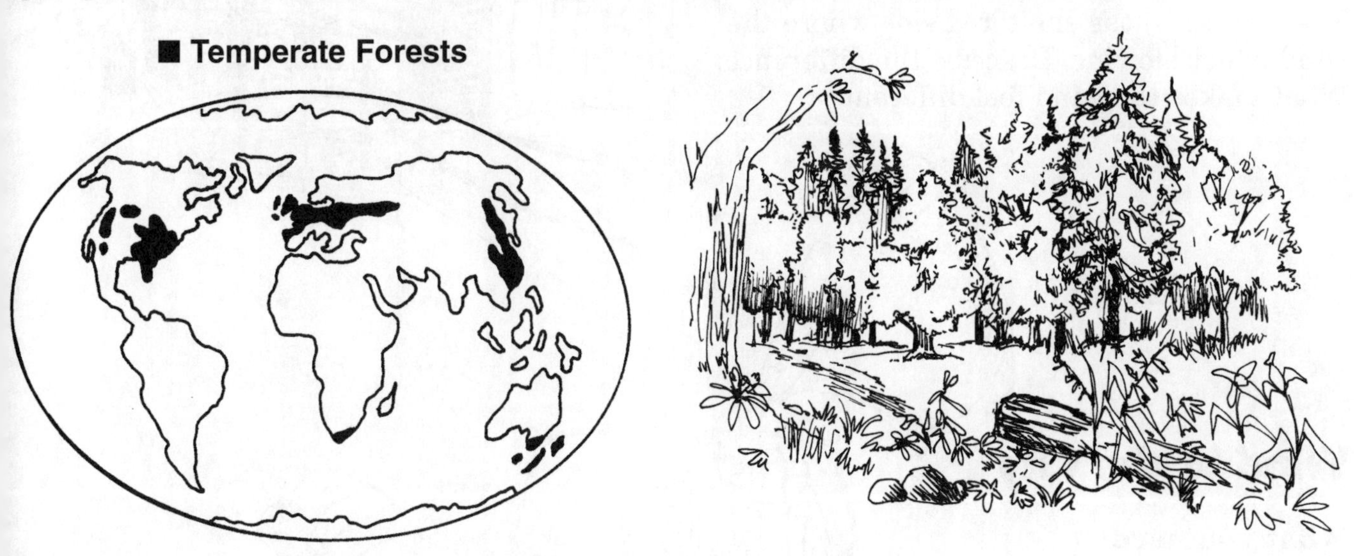

Compare a Deciduous Plant with a Cactus

What you need:

a cactus
a deciduous plant
sand
soil
2 pots
water

What to do:

1. Plant the cactus in a pot of sand. Plant the deciduous plant in soil in the second pot. Don't water either plant.

2. Put the two plants in sunny spots. At the end of two days, observe what happens to each plant.

3. Now water the plants and observe what happens to each plant.

Record your observations in your science journal.

Towns and Cities

Towns and cities are ecosystems, too. There are more than 6 billion people on Earth. The more people there are, the more towns and cities build up. Although the original life may have been driven out, new plants and animals move in to take advantage of the ecosystem.

Buildings act in a city or suburb like trees do in forests. They block the wind or divert the wind to different directions. You can stand by a house or building to feel the wind blowing by. Then stand on the sheltered side where the wind is not blowing. Describe the difference. What makes the wind feel different?

Wind in the City

What you need:

baking pan
black pepper
modeling clay
milk

What to do:

1. Stand modeling clay formed like buildings at one end of the baking pan.

2. Cover the bottom of the baking pan with milk. Sprinkle the pepper on the milk "behind" the buildings.

3. Lift the pan slightly to start the milk flowing. Watch the patterns the pepper makes as the milk moves around the buildings. The flat area in "front" of the buildings simulates a meadow. How is the movement of the pepper different from the buildings to the meadow?

If you live near a forest or seashore, a grassland or lake, a mountain or river, you can study a wide variety of living things in those ecosystems. But even if you don't live near one of those ecosystems, you probably live in or close to a city or town. You can study living things in that ecosystem, too.

Plant Search

What you need:

binoculars (not essential)
an observation sheet or science journal
a pencil
an adult partner

What to do:

1. First, avoid poison ivy, poison sumac, and poison oak. Here is what they look like. Ask your adult partner to make sure you know how to identify poison ivy, poison sumac, and/or poison oak.

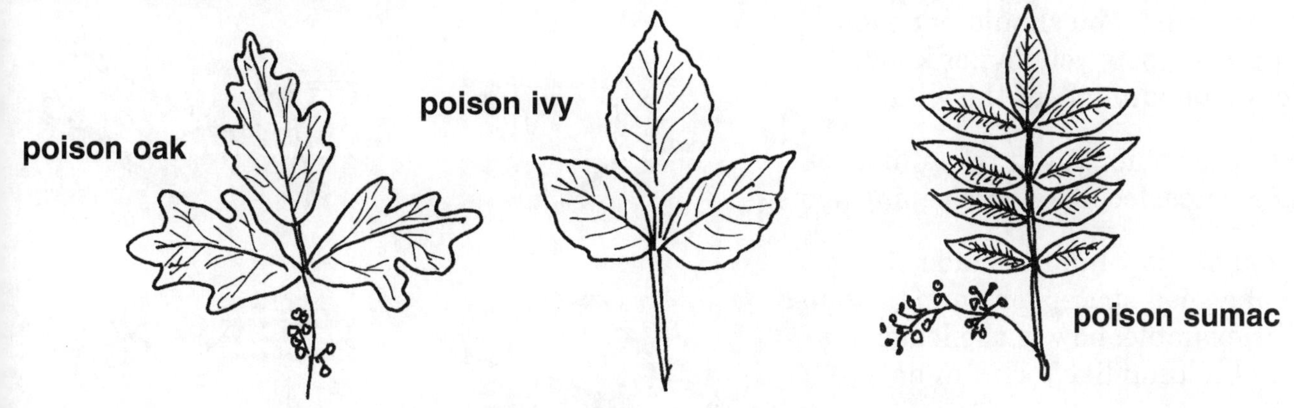

2. Go on a hunt for plants. Look for plants surviving in unlikely places.
3. Record the types of plants you see and draw them.

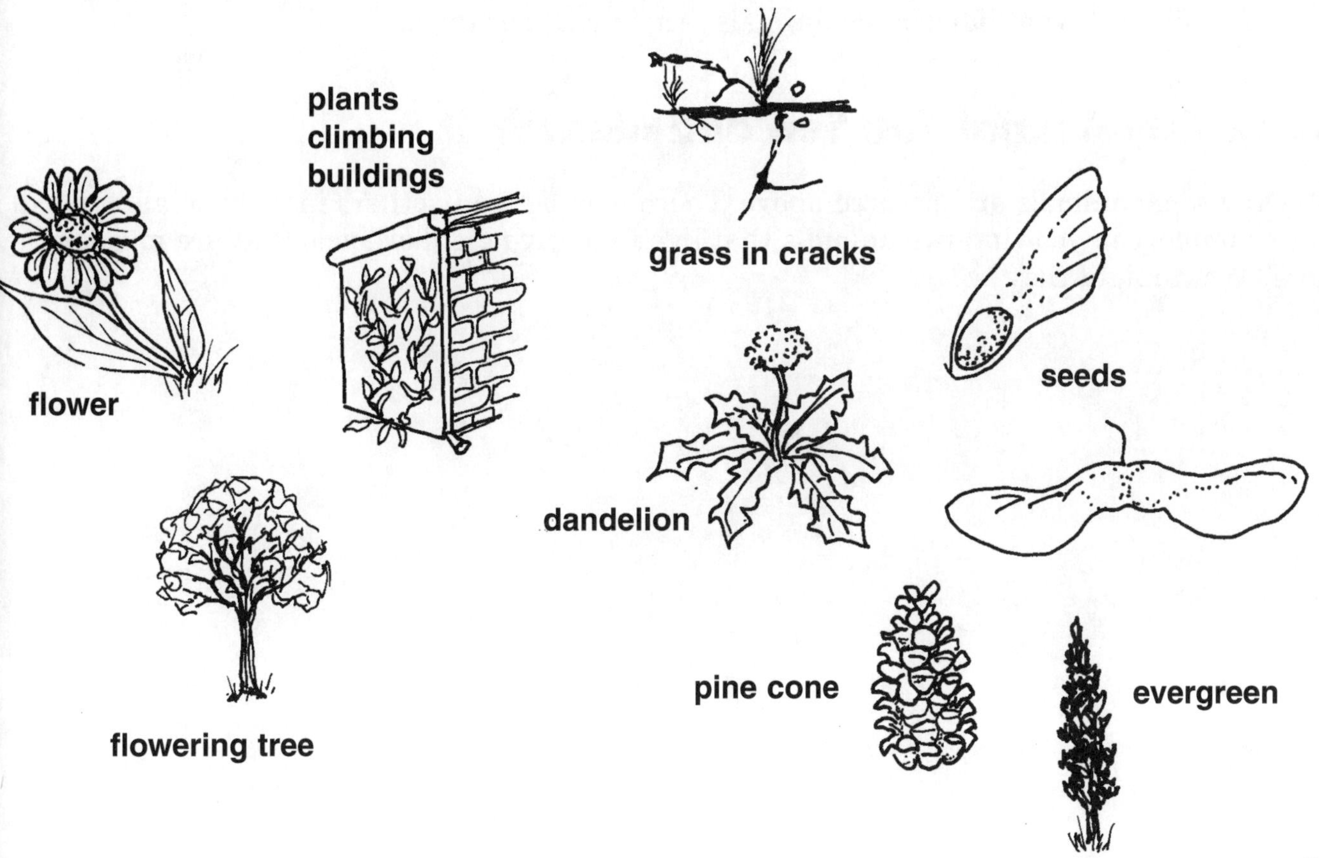

Animal Search

What you need:

binoculars (not essential)
a pencil
an observation sheet or science journal
an adult partner

What to do:

1. First, don't go near strange animals of any kind. You should not even approach pets you do not know, let alone animals in the wild.   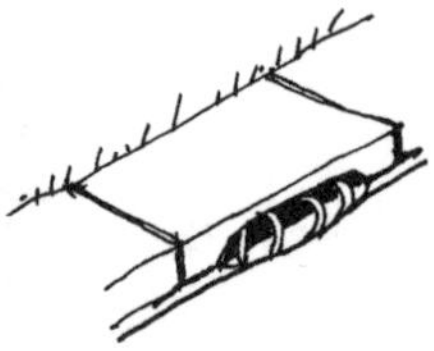
2. Look for places used by animals as nest sites, lookouts, or sleeping spots.
3. Analyze how human-made objects make good spaces for animals to use. For example, how is the ledge of a tall building like a cliff in nature?  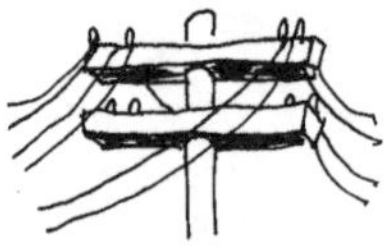 
4. Study all parts of your ecosystem. Remember to look up at utility wires and telephone poles. Look down in the cracks of the pavement and under rocks. As you study your busy ecosystem, record information such as this: What was the best time to observe animals? What food and water sources are available? What dangers do animals face in cities or towns?

Draw Any Animal You See and Identify It

Only some animals are pictured above. There may be many others in your neighborhood. For example, can you find two animals that live in many neighborhoods that are missing here? What might they be?

Does Science Really Affect You?

It's fun to read about discoveries and talk about inventions. But does science really mean anything to you and your everyday life? Think about the most important inventions of the past few centuries. What about the car, the light bulb, the telephone, the television, or computers? Are any of these part of your everyday life? What about antibiotics, vitamins, or new medical equipment?

Now think about the future. What role do you think science will play in the future? Today we face some difficult decisions about things like the environment, hunger, social issues, and the global community. The future—your future—will turn to science more and more to find the solutions to problems and answers to many new questions that are certain to arise. Here are a few problems to start you practicing your problem-solving process.

What you need: **a science journal, some problems to solve, an imagination and inquisitive mind, access to research sources such as a library, experts, a computer, computer software**

What to do:

1. Try to follow the steps on page 9.
2. Select problems that are interesting to you and begin investigating them further. Or come up with problems of your own that you want to explore further.
3. Gather information. Use the information you gained from the experiments in this book. Search for information you need in reference materials, books, videos, and software. Write questions to experts and scientists for information that will help you investigate the problems you select.
4. Make predictions. Try experimenting. Form new questions. Gather more evidence and information.
5. Report your conclusions and your suggestions. Write an article for the newspaper. Prepare an argument for or against an issue. Create solutions to problems and share them within your community.
6. Then write down the new questions your exploration raised and keep exploring.

What About Garbage?

Garbage is everywhere! Look around you. There's probably trash in the wastebaskets, in the streets, in the yards. What kind of garbage do you see around you? Some trash may be paper. Some may be empty cans and bottles. Sometimes people are careless or forget to put garbage where it belongs. People don't always remember that garbage is not very nice to look at or that it pollutes our world. Some garbage, such as broken bottles, can be dangerous.

You probably already know a lot about garbage and why it is a problem in our world today. You can use what you know to begin to figure out some solutions to garbage around you. Think about the trash in your neighborhood. What kind is it? How much is there? Who leaves the trash in your neighborhood? What other kinds of questions do you have about garbage and getting rid of garbage? How can you find some answers to those questions? What are some ways you can help to keep your world free from pollution? Make a list.

What Do You Think About Using Animals for Study?

When *Sputnik II* blasted off a Soviet launch pad on November 3, 1957, it was a historic moment. It was the first satellite to carry a live passenger, a dog named Laika. The first live passenger in an American rocket was Ham, a chimpanzee. Researchers have used and continue to use animals instead of people for research. Each year, researchers use about 17 million animals in laboratory experiments. About 85 percent of research animals are rats and mice. The rest include monkeys, chimpanzees, guinea pigs, horses, cows, rabbits, dogs, and cats. Scientists use these animals to test drugs, cosmetics, and household products, and to study psychological behavior, develop surgical techniques for transplanting body parts, and find cures for diseases. New drugs and new surgical techniques are tested on animals before they are available to humans. Many animals die during these experiments.

Animal experimentation has led to many medical breakthroughs, such as the discovery of insulin (a drug for diabetes), vaccines to prevent polio and smallpox, and treatments for some kinds of cancer, heart disease, and some psychological disorders.

Many people are bitterly divided over the question of the rights of these research animals. Some people believe that the experiments are necessary in order to save human lives. But others believe that the animals are cruelly mistreated and that no one has a right to do research on animals.

How do you feel about this issue? What do you already know about animal research? What are some questions you have that you would like to find answers to in order to make a decision? How can you find more information about animal research in order to decide where you stand on this issue? Use what you find out to write and present a persuasive talk explaining your position about the use of animals in medical, product, and/or space research.

Can Our Soil Be Polluted?

Water and air can be polluted. But can soil also be polluted? After all, land is just land, isn't it? But we need our soil for many reasons. As long as there is enough land to grow food, there should be no problem, right? Well, not exactly.

A thin layer of fertile soil covers much of Earth's land surface. This soil contains nutrients that plants need to grow. If the soil becomes polluted, plants may not grow well.

Many farmers use chemical fertilizers and pesticides. The fertilizers make the crops grow better. The pesticides kill insects and weeds that might damage the crops. Both fertilizers and pesticides help farmers grow more crops and earn more money. Unfortunately, some of these chemicals can also damage the nutrients in the soil. Certain types of crops use up large amounts of nutrients, too. If the nutrients are destroyed, the soil is less able to grow new crops. Sometimes the fertile layer of soil wears away through erosion. When trees and grasses are removed, the wind and rain can wash away fertile soil. Careless farming methods can also cause erosion.

How can we avoid polluting the land or losing it to erosion? What do you think? What information do you already know? What questions do you have? What is some information that might help you make some decisions about this issue? What do you need to know to begin developing some ideas on ways to help avoid pollution and unnecessary erosion? Develop several solutions that you think could solve the problem.

Graphic Organizers

Graphic organizers do exactly what their name suggests. These graphs and charts help you to organize information for clarity, to better understand concepts, or to organize information for reference later. Use these graphic organizers to help you organize what you are learning. Each type of graphic organizer shown contains an example of how it can be used.

Activity: Establishing Criteria and Summarizing

Problem-Solution Frame

Example: Health Choices

Problem Box

What is the problem? I have to make choices every day that affect my health.

Why is it a problem? If I make poor choices, my health can be affected.

Solution Box

Solutions	Results
1. I can be aware of my physical and emotional health.	1. I will realize ways my choices affect my health.
2. I can make safe choices when I'm playing outside.	2. I will have more control over my own safety.
3. I can get exercise every day and get a good amount of sleep.	3. I will feel better and be healthier.

End Result Box

If I am aware of my health choices, then I will have more control over my own life.

Decision-Making Chart *Example: Choosing a bike*

		Alternatives		
	Weights	Brand X	Brand Y	Brand Z
ten-speed	3	3		3
lightweight	1	1		1
costs less than $300	2	2	2	
green or blue	1		1	1
Total		6	3	5

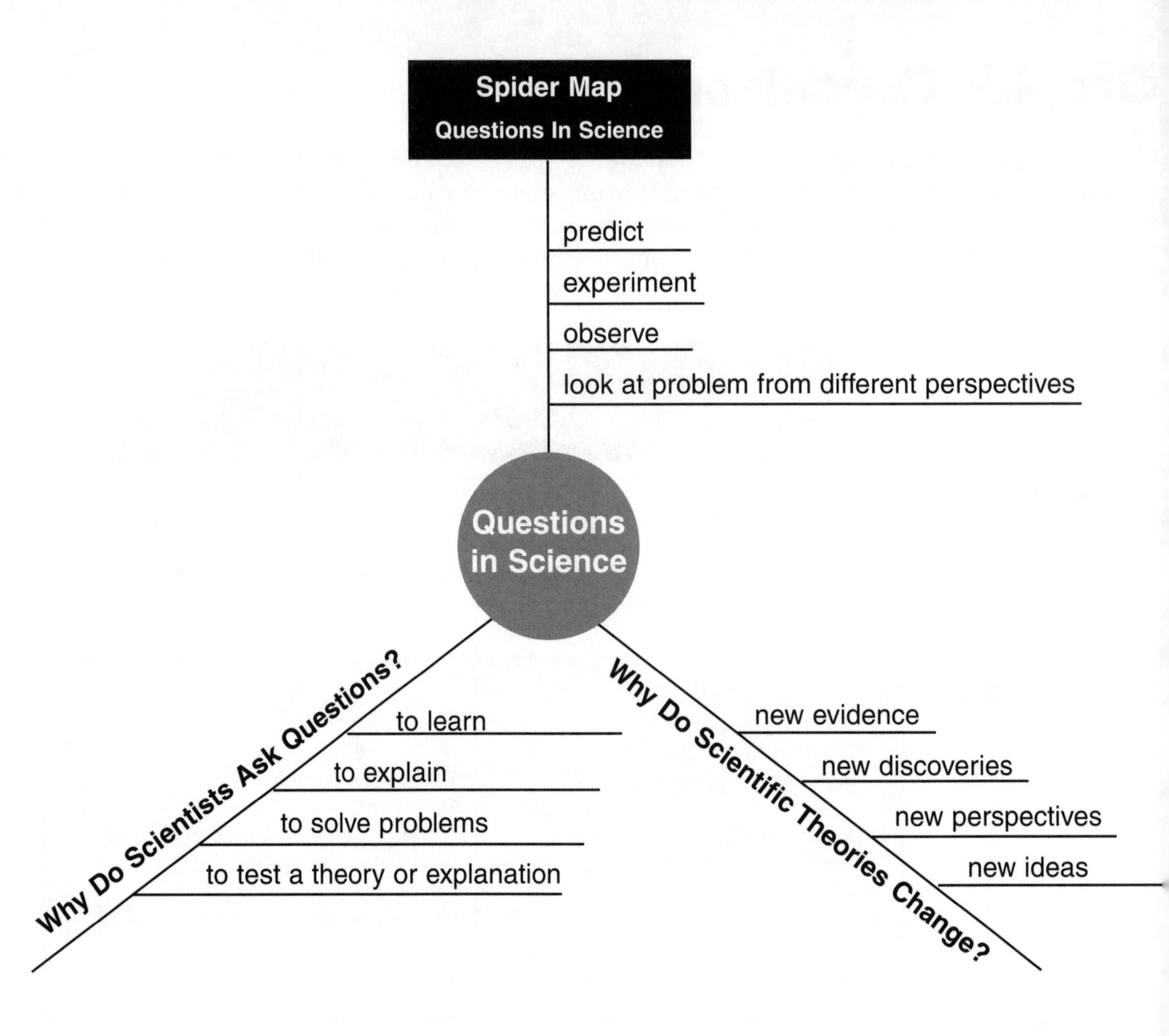
Spider Map
Questions In Science
predict
experiment
observe
look at problem from different perspectives
Questions in Science
Why Do Scientists Ask Questions?
to learn
to explain
to solve problems
to test a theory or explanation
Why Do Scientific Theories Change?
new evidence
new discoveries
new perspectives
new ideas

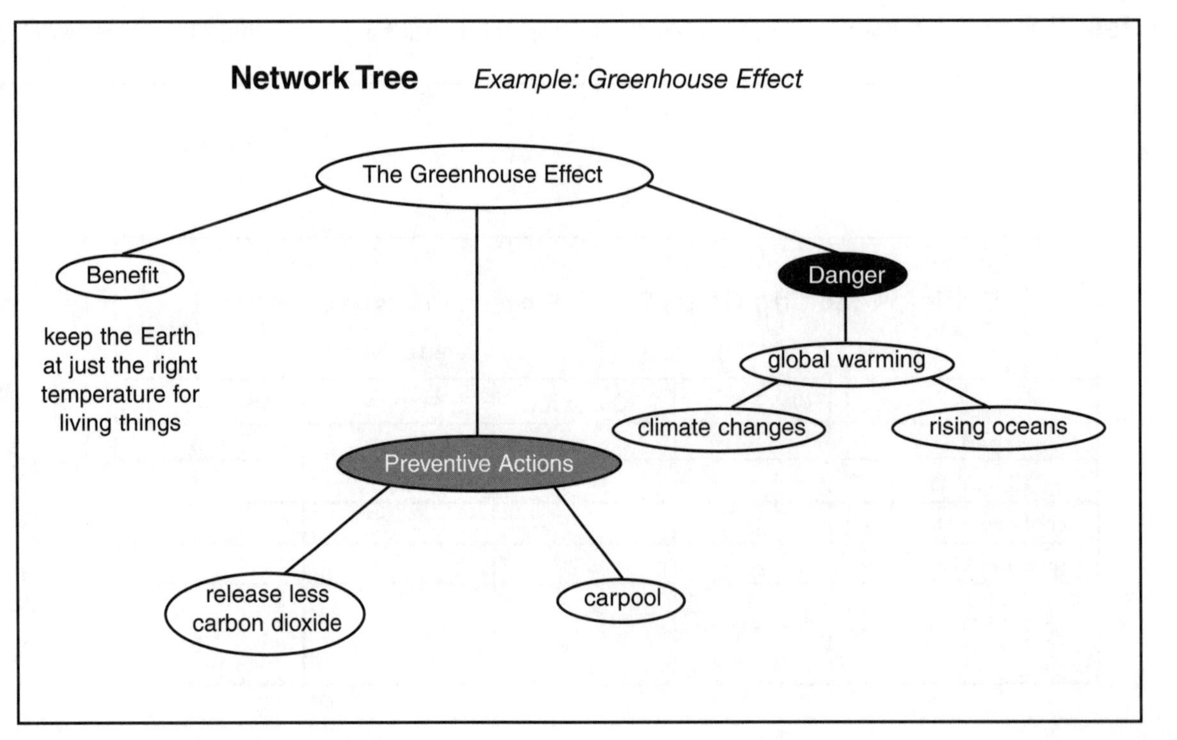
Network Tree
Example: Greenhouse Effect
The Greenhouse Effect
Benefit
keep the Earth at just the right temperature for living things
Danger
global warming
climate changes
rising oceans
Preventive Actions
release less carbon dioxide
carpool

Fishbone Map *Example: Polluted Water*

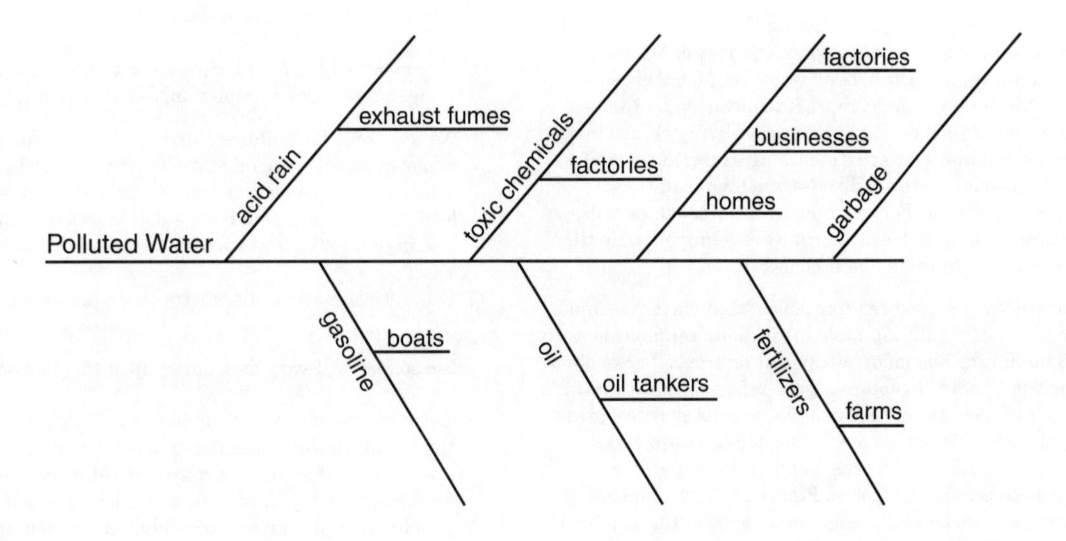

Comparison Matrix *Example: Earth/Mars*		
Feature	**Earth**	**Mars**
Atmosphere	mostly mixture of oxygen, nitrogen, and carbon dioxide	mainly carbon dioxide
Water	abundant amounts appear as liquid, solid, and vapor	may once have existed on the surface and may exist underground and in caps at both poles; very slight traces of water vapor in atmosphere
Temperature	–89°C to 58°C	–124° to –31°C
Safe Environments	a variety of environments provide protection	unknown

Answers

Remember, accept any reasonable response. Children should be encouraged to be creative. If they come up with an interesting new response, check it out together.

Page

9. Start a science journal to use while working through this book.

10. Accept any reasonable responses. Living things respond, need water and food to live, reproduce, grow. Plants and animals have movement and response, need water and food, grow. Non-living things do not grow, reproduce, or move by themselves, do not need food or water. Some non-living things might include rocks, the house or apartment, furniture, automobiles, pencils, pens, books. However, discuss what some of these things were at one time. For example books, wooden pencils, and furniture came from trees, which were living at some time. This is a good time to think about conservation.

    Questions will vary. A good reference book that classifies and shows the kinds of plants you have in your environment is a big help. The library has many wonderful reference books about plants. You might want to obtain a good reference book for animals, as well. Accept any reasonable responses: roses, grass, tulips, maple trees, dogwood trees, elms, pines, clover, oaks, palms, moss, ferns, daisies, cactus, heather, cabbage, carrots, legumes, irises, parsley, foxgloves. Plants can have flowers, petals, seeds, leaves, trunks, stems, roots, cells, anthers, pollen, ovules, ovaries. Plants need water, light, nutrients, carbon dioxide. Some plants lose their leaves in fall; some grow very slowly in the winter; some flower in the spring, then flowers turn to fruits in the summer or as the growing season becomes warmer; some change little during the different seasons. Seeds are "baby plants." They are produced in flowers, in flowering plants, and in cones in conifers. They are produced when the pollen and the ovule or egg combine. They grow into new plants of the same kind as the parent plant.

11. *Diverse* means "differing from one another." Have a good dictionary available while working in this book.

    Signs of plants can include such things as fallen leaves, grass clippings, seeds, twigs. Weeds are usually pretty easy to determine. Most flowers around houses and other buildings are domestic, or planted there. Discuss that it can be difficult to tell the difference between some weeds and some domestic flowers. Some flowering wild plants are very pretty. It is important to show the characteristics of poisonous plants such as poison sumac, poison oak, and poison ivy. Also discuss the rule to never eat anything growing wild. It takes training to learn which wild fruits, nuts, and mushrooms are not poisonous and which ones are.

12. A hypothesis is "an assumption" or an inference based upon prior knowledge or interpretation of facts or situations. Accept any reasonable hypotheses and applaud creativity.

    Plants need nutrients in soil. They also need carbon, most of which they get from carbon dioxide in the air. Make sure plants get the amount of sunlight they require. Make sure they get the amount of water they require. Make sure the soil is rich with nutrients.

13. $\frac{1}{100}$th of an inch is smaller than a period on page 13. Show $\frac{1}{16}$ of an inch and extrapolate from there.

    "Blooming" is a British adjective that means "wonderful" or "excellent."

Page

14. Ovules (eggs) and anthers may be hard to find in some flowers. Pollen usually is fairly easy to find. It is the colored dustlike material that gets on hands or clothes.

15. Responses will vary depending upon the plants selected, but they should have roots, stems, and leaves. Some may have flowers like the plant in the picture.

16. To conserve means to keep safe or to protect something from danger, destruction, depletion, or extinction.

    We use trees for building; instruments and furniture; paper and paper products; pencils and other writing tools; transportation such as ships, canoes, or boats; tree houses to play in; to hold food such as ice-cream bars, salad bowls, or chopsticks or cooking utensils; toys such as wagons, doll houses, and blocks.

17. Some seeds, such as strawberry seeds, are on the outside of the fruit.

    Responses will vary depending upon the food selected.

18. Eggs and butter come from animals. People-made foods include the bread, popcorn, and french fries. The ingredients of bread are some kind of grain, such as wheat or rye; eggs from animals; possibly milk from animals; some salt, a mineral; pepper from plants; and possibly butter from animals or margarine from plants. The popcorn probably includes butter from animals; salt, a mineral; and is in a paper container. The french fries probably were fried in animal or vegetable oil, probably have salt on them, and, like the popcorn, are in a paper package. The other foods grow as they are on plants.

19. Make sure the simulated maple tree seed has a small paper clip on the bottom for weight as a seed does.

20. Plants need plenty of sunlight and water.

21. Remember, a good reference book that classifies and shows animals is a big help as you begin exploring. Animals need food, water, and air to breathe. Use the dictionary. *Satisfy* means "to please, to fulfill the needs of." Birds and bats can fly. Not all birds can fly. For example, penguins cannot fly. Flying squirrels do not actually fly. They use the loose skin attached to their front legs to catch the wind like a kite and sail. Also, flying fish cannot fly. They simply leap high out of the water for short distances. Fish, sharks, whales, porpoises, and dolphins can only swim. Giraffes, rhinos, and elephants can only walk, crawl, or run. Many animals can swim if they have to, such as cats and dogs, but that's not their normal method of transportation. People can crawl, walk, climb, and swim. So can apes, squirrels, opossums, monkeys, chimpanzees, and cats. Many animals use their paws to help capture food. Birds use their beaks to find food. One animal that may not come to mind is a squid. Squid also have what is called a beak with which they hunt food. So does a duck-billed platypus. Birds have good eyesight for finding food from above. Many animals, such as cats, dogs, apes and chimpanzees, have very good eyesight. People have good eyesight, as well. Bats use their ears to "see." Many fish and marine animals also use "echolocation" to search for food and help navigate.

Page

22. Usually, the hardest direction is directly in front or directly behind because both ears are hearing about the same sound. A bat cannot see. They use sound vibrations to locate food and to navigate.

23. The sounds that reflect off solid objects are different from the sound that reflects off moving objects. The sound reflects differently depending on the size of the object, as well. To a bat, the reflected sound from a tree would be different from the reflected sound of a piece of fruit, which might be his/her dinner. Besides bats, some water animals use something called "echolocation" as well to find food and navigate. Dolphins are expert at using echolocation.People in the fishing industry use echolocation to find schools of fish. Oceanographers and marine biologists use echolocation as they study bodies of water and the life that lives there. Earth scientists use a kind of echolocation to study the earth and to study earthquakes.

24. Some frogs, toads, lizards, and snakes use their tongues. Put a little curled piece of masking tape on the tip of the party whistle. Applaud creativity.

A straw can be used like a hummingbird's beak. The glass is like the flower. The milk is like the nectar.

25. Gelatin will be very difficult to pick up with chopsticks or with the ice-cream bar sticks. Also, small objects such as uncooked rice will probably be difficult to pick up. The ice-cream bar sticks will probably pick up some items easier because they are bigger and can be used like a scoop as well. You could break up the cookie with the chopsticks or pick the cookie up with the ice-cream bar sticks and take a bite. Finding a way to eat the cookie will probably not cause a problem. Applaud creativity. Fish, whales, dolphins, porpoises, and most water animals use their mouths. Hoofed animals such as giraffes, cattle, horses, and sheep use their mouths to graze. Many animals use their mouths in one way or another to find and capture food.

26. In this context, and/or means an animal can possibly swim, climb, and fly, but can possibly only swim or fly. A monkey can run, walk, and crawl. A duck can swim, walk, and even run. Discuss swimming. Can the student swim?

Applaud creativity.

27. The skeletal system is the firm structure that supports the body. It can be an exoskeleton in insects, arachnids, and crustaceans. In sharks and some other cartilaginous fish, the skeletal system is composed of cartilage, not hard bone. In birds, reptiles, amphibians, fish, and mammals, it is the system of bones that support the body and to which the muscles are attached.

Responses will vary. Accept any reasonable response, such as muscles help animals move, lift things, pump blood. Muscles also act as protection and support. No skeleton at all—Portuguese man-of-war, earthworm. Exoskeleton—crab, insect, spider. Internal skeleton—bear, alligator, bird, fish.

28. The index fingers will curve toward each other slightly when relaxed. The tendons "tend" to pull the muscles because of the "tension." Notice the similarity in words.

29. Pushing hard against the door frame for a period of time forces the muscles to become accustomed to that position. Then when released, the muscles and tendons will remain adjusted to that position for a short time, causing the arm to rise, seemingly on its own.

Page

30. book = load/adult; muscles = force/child; elbow = fulcrum/triangle object

If you attach another string from the lower part of the simulated upper arm to the "wrist" area of the simulated lower arm, you can simulate a muscle in the lower arm and show how that muscle also helps the arm move.

31. The bones will bend only one way. They do not bend back without breaking. The white-colored substance near each joint is the cartilage. The cartilage protects the joints and makes it easier for them to move. Infer means to make an "educated" guess based upon prior knowledge. Always remember to check or verify inferences just like hypotheses.

Friction is the resistance between moving objects where they contact each other.

32. The jaw works like a pair of tweezers. The food is the load. The jaw joint is the fulcrum. The muscles are the force. It also works like a nutcracker when the food is close to the back teeth. Still, the joint is the fulcrum, the food the load, and the muscles the force.

33. An ecosystem is a distinct area in the biosphere that contains living organisms such as a lake, a forest, a desert, a grassland, river, wetland, ocean, tundra, mountain, seashore, polar land, tropical rain forest, even a city or town.

summer, winter, spring, fall or autumn

34. Responses will vary, depending on location. Reference materials such as libraries, software, and books should be available. Fall originated from the fact that leaves in some areas do "fall" off the deciduous trees.

This is a good place to discuss the issues of health and safety as they relate to weather.

37. They should weigh the same since they were cut the same size. The only difference is that the simulated sleet is crumpled. The surface area of the simulated snowflake is larger, so there is more space for more air resistance as it falls. In the same way as the simulated snowflake, the parachute has a large area for resistance, so the air pushes up into the parachute as the parachuter falls through air.

38. Be careful not to slap the ruler too hard or too fast. It could break from the force and the pressure, even though that may be hard to believe.

39. As with all matter, air does have weight and takes up space.

40. Responses will vary depending upon location.

If the pine cone does not close, try putting it closer to the warm water.

41. Anyone who lived or lives on a farm has probably observed animal responses. Some animals are very sensitive to their environment and changes in the environment.

42. The lard should insulate the hand from the cold. The hand in the "lard" glove will not feel as cold as the hand placed directly in the cold water. A seal's, whale's, dolphin's, or porpoise's layer of fat (blubber) insulates the animal from the icy water in which it swims. Layers of fat on land animals, such as bears, insulate along with their heavy furs. Some animals, such as bears, build up their fat throughout the spring, summer, and

Page

42. fall for several reasons. The extra fat helps them survive through hibernation and also insulates them from the cold winters.

43. The hand in the warm water is gaining heat from the water. The hand in the cold water is losing heat to the cold water. When you place both hands in the room temperature water in pan #3, the hand that was in the cold water senses the room temperature water as warm because it is absorbing heat from the water, whereas the hand that was in the warm water feels cool because it is losing heat to the water. You need to be very observant to feel the differences. This experiment shows the reason your hands don't make a very good thermometer—they each feel the same room temperature differently. The hand that was in the cold water feels the room temperature water as warm but the hand that was in the warm water feels the room temperature water as cool.

44. Inventions will vary depending upon materials available. Applaud creativity. The invention should have some kind of insulation between the inside of the container and the outside environment.

46. The warmth of the hand warmed the air in the jar. The warm air molecules move more and the warm air rises trying to get out of the jar. The work the warm air does is to raise the lid off the jar to escape.

    If the seal is tight around the straw, the warm air should put pressure on the water, pushing down harder on the water. Since the air can't rise and get out, the cooler air will sink instead. When it pushes down on the water, the only place the water can move is up the straw. So the water level in the straw should rise. If the water in the straw does not move, then the seal around the straw is not tight enough. Try again.

47. The heat lifted the lid. The heat raised the level of water in the straw. The sun's energy is doing the same kind of work a microwave oven or traditional oven does, heating the chocolate enough to melt it. This is a good place to discuss the use of the sun and solar energy as a renewable energy source.

48. Responses will vary depending upon the location and the amount of sun coming through the window. Be very careful when trying to compare the effect of the light with the sun. Discuss how important it is never to look directly at the sun.

49. When the lamp is behind the milk pitcher and you look through the milk to the light, the color will be more orangish than white or bluish. But when you look at the milk/water from a right angle to the lamp, the color will be more bluish or gray-white. More light should shine through the glass than through the pitcher, but the color comparison should be the same.

50. A rainbow can be seen in nature. A natural rainbow results from the sun's rays reflecting through raindrops or mist. Mist near waterfalls also can produce a rainbow of colors.

51. It is important to discuss the fact that during an eclipse you should never look directly at the sun. The rays surrounding the moon in a full eclipse are still very dangerous. Total solar eclipses do not happen very often. One took place so people in parts of North America could see it in 1999. Another one will take place in 2008.

    Using your finger, you can block out large objects in the distance. Relative sizes are affected by distance. Standing next to a train, the train seems very large, but from an airplane 10,000 feet in the air, it is very small.

Page

52. It is important to understand that simulations are only that, simulations. It is almost impossible to demonstrate on a piece of paper the relative sizes of the planets. Also, their orbits are not perfectly round. They are oval in shape. In addition, the distances between planets are difficult to simulate since there really is not enough space on a piece of paper.

    Sometimes it is more understandable to hold objects that better show the relative sizes of planets than to look at pictures. Although the foods listed here are not perfect simulations of the relative size of each planet, they do give a better appreciation for how small and how large certain planets are compared with others.

53. Remember, the orbits are not round. Again, this is a simulation. The amount of time it takes to orbit the sun really gives a better understanding of distances than looking at a picture or than reading the distances in feet, miles, or light-years. Mercury and Venus will make more than one orbit. Earth will make one orbit since this is based on Earth years. The rest of the planets will make less than one orbit.

    Mars = 1 year and 11 months, Jupiter = 12 Earth years to orbit the sun, Saturn = 30 Earth years to orbit the sun, Uranus = 84 Earth years to orbit the sun, Neptune = 165 Earth years to orbit the sun

54. There are 11 zeros in 100,000 million = 100,000,000,000. Applaud creativity.

56. There are 5,280 feet in one mile. Scientists estimate Earth was formed 5,000 million years ago. So 52,800,000 years is a relatively short time.

    Sand will act similarly to rocks and dirt carried in real glaciers. It will cause indentations in the surface of the clay. As the ice cube (glacier) moves, it melts in the crater it is forming. The melting eventually will leave a lake.

57. The water pollution will flow both into the lake and into the ground around the lake, so pollution will be absorbed by living things both in lakes and living things that live around lakes.

58. Much of the "pollution" settled in the higher pan. Some of the "pollution" was absorbed by the cloth filter. That may be evident by a discoloration of the cloth. The food coloring will probably be absorbed by the cloth. The food coloring will probably be lessening in intensity the farther down the cloth filter it goes. However, some food coloring could reach the lower pan and may not be seen because it is diluted. But it still may be there.

    There are many causes of water pollution, such as fertilizers, soaps, insecticides, and waste products. Applaud creativity in ways to slow pollution—starting at home by using less soap, and fewer or no fertilizers and/or insecticides to outside the home by sending letters and articles to newspapers or to factories that are known polluters.

59. Even a weak acid like vinegar will begin breaking down the limestone in chalk. The bubbles in the vinegar solution are carbon dioxide produced in the reaction between vinegar and limestone. This is similar to what happens in nature. Look at some pictures of very old buildings, such as the Coliseum in Rome or the Parthenon in Athens, the pyramids, the Sphinx, and so on. They are eroding faster all the time due to the acid in air pollution. Rain and weather have eroded these structures naturally, but the acid pollution in the air combined with rain is speeding the process of erosion. Look for statues and buildings closer to home that are showing signs of erosion.

Page

59. The acid in the vinegar dissolves the calcium in the eggshell. Vinegar is a weak acid. Acids produced by the combination of sulfur dioxide from burning and rainwater produce a much stronger acid in acid rain, sulfuric acid.

60. The highest mountain in the world is Mount Everest in Nepal, north of India. It is 29,035 feet high at its peak. Responses will vary depending on location. *Plate* is the part of the word that indicates the flatness of plateaus.

61. Responses will vary depending on location. Plans can range from an experiment as simple as testing how much water the soil will hold to actually having the soil tested. If there is a park administration or an EPA office in your area, they are good sources of information. The library is also a good source of information about your area. Another excellent source for information about soil is any farmer or farm organization in your area or your local plant or gardening store.

    A sphere is a round object like the earth. *Bio* means life.

62. The ice cubes simulate the frozen polar regions. As they melt, the water drains into the water surrounding the simulated island. Eventually, the simulated island will be covered with water. The simulated island simulates both islands and shorelines that could be covered by melting polar ice.

63. The wind moves the surface waves. Temperature and land movement in and under the ocean are other causes of currents. The wind results from warm and cold air patterns, not from a giant person blowing on the surface as in the simulation. Remember the experiments conducted on warming and cooling air and what happens when air is warmed.

64. Responses will vary depending upon the location. "Teeming with life" means an ecosystem has many different kinds of living organisms. It is abundant with life.

65. Cacti is the plural of cactus. This might be a good place to look at unusual plurals. There are a lot of them in the names of animals and plants.

    The spaghetti acts as a barrier for the sand. Some sand will get through, but less than when there is no protection. The spaghetti simulates trees, which are nonexistent on large stretches of desert. Many beaches and parks have fences maintaining the sand in an attempt to slow erosion. Some beaches build dunes to contain the sand, and some beaches have natural dunes, such as the sand dunes at Nags Head, North Carolina.

Page

66. Tropical rain forests are normally considered those that are near the equator. However, other temperate forests are considered a kind of rain forest as well. The forests in the Northwest, throughout Washington and Oregon, are temperate forest but are, to some ecologists, also considered rain forests, just not tropical rain forests. This is an interesting fact. What is the biggest export from Brazil? Answer: oxygen.

    The jar that was placed in the warm sunny spot simulates the rain forest. In tropical rain forests, the remains of living organisms decay very quickly and plants absorb the nutrients just as quickly. Because plants grow throughout the year, little decaying material accumulates.

67. Temperate forests are usually seasonal. In other words, unlike tropical rain forests that maintain a similar temperature and in which plants grow at the same rate throughout most of the year, most temperate forests go through slow growing seasons and often dry seasons.

    The cactus will probably not change at all without water for several days. As a matter of fact, cacti should not be overwatered. However, the deciduous plant will begin to wilt without water. The deciduous plant requires more water than the cactus. However, like the cactus, it too can be overwatered.

68. The building actually blocks the wind. Sometimes the wind blocked by a building will rush even faster or with more strength through the streets because the buildings allow less area for the wind to pass through, thus condensing the wind into a smaller passageway.

    The simulated buildings reroute the direction the simulated "wind" (milk) is traveling. When the simulated "wind" (milk) reaches the open space, it disperses over a larger area. The pepper is simulated particles in the air—dust, dirt, pollution. You can see how it collects in corners of simulated "buildings."

70. You can find animals almost everywhere. Some places you just have to look a little harder or look for signs of animals such as chewed leaves, egg cases, webs, paths, chewed seeds. Insects are everywhere, and there are very few places where there aren't at least a few cats or dogs.

    Remember, a good animal reference book is very helpful. Chances are you forgot dogs or cats. Also, what about you?

Notes